T0205395

Intelligent Systems Reference Library

Volume 121

About this Series

The aim of this series is to publish a Reference Library, including novel advances and developments in all aspects of Intelligent Systems in an easily accessible and well structured form. The series includes reference works, handbooks, compendia, textbooks, well-structured monographs, dictionaries, and encyclopedias. It contains well integrated knowledge and current information in the field of Intelligent Systems. The series covers the theory, applications, and design methods of Intelligent Systems. Virtually all disciplines such as engineering, computer science, avionics, business, e-commerce, environment, healthcare, physics and life science are included.

More information about this series at http://www.springer.com/series/8578

Chengjun Liu
Editor

Recent Advances in Intelligent Image Search and Video Retrieval

 Springer

Editor
Chengjun Liu
Department of Computer Science
New Jersey Institute of Technology
University Heights, Newark, NJ
USA

ISSN 1868-4394   ISSN 1868-4408   (electronic)
Intelligent Systems Reference Library
ISBN 978-3-319-84816-7   ISBN 978-3-319-52081-0   (eBook)
DOI 10.1007/978-3-319-52081-0

Printed on acid-free paper

This Springer imprint is published by Springer Nature
The registered company is Springer International Publishing AG
The registered company address is: Gewerbestrasse 11, 6330 Cham, Switzerland

Preface

Digital images and videos are proliferating at an amazing speed in science, engineering and technology, media, and entertainment. With the huge accumulation of such data, the keyword search or manual annotation scheme may no longer meet the practical demand for retrieving the relevant contents in images and videos. Intelligent image search and video retrieval thus have broad applications in, for example, smart cities such as intelligent traffic monitoring systems, surveillance systems and security, and Internet of things such as face search in social media and on web portals.

This book first reviews the major feature representation and extraction methods and the effective learning and recognition approaches that have broad applications in intelligent image search and video retrieval. It then presents novel methods, such as the improved soft-assignment coding, the inheritable color space (InCS) and the generalized InCS framework, the sparse kernel manifold learner method, the efficient support vector machine (eSVM), and the SIFT features in multiple color spaces. This book finally presents clothing analysis for subject identification and retrieval, and performance evaluation of video analytics for traffic monitoring.

Specifically, this book includes the following chapters. Chapter 1 reviews the following representative feature representation and extraction methods: the spatial pyramid matching (SPM), the soft-assignment coding (SAC), the Fisher vector coding, the sparse coding and its variants, the local binary pattern (LBP), the feature LBP (FLBP), the local quaternary patterns (LQP), the feature LQP (FLQP), the scale-invariant feature transform (SIFT), and the SIFT variants. These methods have broad applications in intelligent image search and video retrieval. Chapter 2 first discusses some popular deep learning methods, such as the feedforward deep neural networks, the deep autoencoders, the convolutional neural networks, and the Deep Boltzmann Machine (DBM). It then reviews one of the popular machine learning methods, namely the support vector machine (SVM): the linear SVM, the soft-margin SVM, the nonlinear SVM, the simplified SVM, the efficient SVM (eSVM), as well as the applications of SVM to image search and video retrieval. Finally, it briefly addresses other popular kernel methods and new similarity measures. Chapter 3 first analyzes the soft-assignment coding or SAC method from the perspective of kernel density estimation and then presents an improved SAC (ISAC) method by introducing two enhancements—the thresholding normalized visual word plausibility and the power transformation method. The ISAC method is able to enhance the image classification performance when compared to the SAC method while keeping its simplicity and efficiency. Chapter 4 proposes a novel inheritable color space or InCS and its two important properties: the decorrelation property and the robustness to illumination variations property. It further presents a new generalized InCS framework to extend the InCS from the pixel level to the feature level for improving the verification performance as well as the robustness to illumination variations. Chapter 5 introduces a new sparse kernel manifold learner (SKML) method for learning a discriminative sparse representation by considering the local manifold structure and the label information based on the marginal Fisher criterion. The objective of the SKML method is to minimize the intraclass scatter and maximize the interclass separability that are defined based on the sparse criterion. This chapter further presents various novel features, such as a new DAISY Fisher vector (D-FV) feature, a WLD-SIFT Fisher vector (WS-FV) feature, an innovative fused Fisher vector (FFV) feature, and a novel fused color Fisher vector (FCFV) feature. Chapter 6 contributes a new efficient support vector machine or eSVM for image search and video retrieval in general and accurate and efficient eye search in particular. The eSVM is able to significantly reduce the number of support vectors and improve the computational efficiency without sacrificing the generalization performance. The eSVM method is an efficient and general learning and recognition method, and hence, it can be broadly applied to various tasks in intelligent image search and video retrieval. Chapter 7 presents new descriptors— the Color SIFT Fusion (CSF), the Color Grayscale SIFT Fusion (CGSF), and the CGSF+PHOG descriptors—by integrating the oRGB-SIFT descriptor with other color SIFT features. These new descriptors are effective for image classification with special applications to image search and video retrieval. Chapter 8 derives a perceptive guide for feature subset selection and enhanced performance through the extended analyses of the soft clothing attributes and the studies on the clothing feature space via detailed analysis and empirical investigation of the capabilities of soft biometrics using clothing attributes in human identification and retrieval. This chapter also offers a methodology framework for soft clothing biometrics derivation and their performance evaluation. Chapter 9 discusses a pilot study that is designed and conducted to evaluate the accuracy of a video analytic product by integrating it with the closed-circuit television (CCTV) cameras deployed on highways. This pilot study is designed to evaluate the accuracy of video analytics in detecting traffic incidents and collecting traffic counts.

Newark, NJ, USA   Chengjun Liu
November 2016

Contents

Contributors

Shuo Chen The Neat Company, Philadelphia, PA, USA

Branislav Dimitrijevic New Jersey Institute of Technology, Newark, NJ, USA

Slobodan Gutesa New Jersey Institute of Technology, Newark, NJ, USA

Emad Sami Jaha King Abdulaziz University, Jeddah, Saudi Arabia

Kitae Kim New Jersey Institute of Technology, Newark, NJ, USA

Yukhe Lavinia California State University, Fullerton, CA, USA

Joyoung Lee New Jersey Institute of Technology, Newark, NJ, USA

Chengjun Liu New Jersey Institute of Technology, Newark, NJ, USA

Qingfeng Liu New Jersey Institute of Technology, Newark, NJ, USA

Wasif Mirza New Jersey Department of Transportation, Ewing Township, NJ, USA

Mark S. Nixon University of Southampton, Southampton, UK

Ajit Puthenputhussery New Jersey Institute of Technology, Newark, NJ, USA

Jeevanjot Singh New Jersey Department of Transportation, Ewing Township, NJ, USA

Lazar Spasovic New Jersey Institute of Technology, Newark, NJ, USA

Abhishek Verma California State University, Fullerton, CA, USA

Acronyms

ANN	Artificial Neural Networks
ANOVA	Analysis of Variance
AUC	Area under the Curve
BBSO	Big Bear Solar Observatory
BDF	Bayesian Discriminating Features
CGSF	Color Grayscale SIFT Fusion
CLBP	Completed Local Binary Pattern
CMC	Cumulative Match Characteristic
CN	Color Names
CNN	Convolutional Neural Networks
CSF	Color SIFT Fusion
CSIFT	Colored SIFT
CSM	Color Similarity Measure
DBM	Deep Boltzmann Machine
DFB	Distance From Boundary
D-FV	DAISY Fisher Vector
DHFVC	Deep Hierarchical Visual Feature Coding
D-KSVD	Discriminative K-Singular Value Decomposition
DoG	Difference of Gaussian
DR	Detection Rate
DSP	Domain Size Pooling
EER	Equal Error Rate
EFM	Enhanced Fisher Model
ERM	Empirical Risk Minimization
eSVM	efficient Support Vector Machine
FAIR-SURF	Fully Affine Invariant SURF
FAR	False Alarm Rate
FCFV	Fused Color Fisher Vector
FDDL	Fisher Discrimination Dictionary Learning
FFV	Fused Fisher Vector

FLBP   Feature Local Binary Patterns
FLD   Fisher Linear Discriminant
FLQP   Feature Local Quaternary Patterns
FRGC   Face Recognition Grand Challenge
GInCS   Generalized Inheritable Color Space
GMM   Gaussian Mixture Model
ILSVRC   ImageNet Large Scale Visual Recognition Challenge
InCS   Inheritable Color Space
ISAC   Improved Soft-Assignment Coding
JDL   Joint Dictionary Learning
JPEG   Joint Photogrphic Experts Group
KKT   Karush-Kuhn-Tucker
LaplacianSC   Laplacian Sparse Coding
LBP   Local Binary Pattern
LC-KSVD   Label Consistent K-Singular Value Decomposition
LLC   Locality-constrained Linear Coding
LoG   Laplacian of Gaussian
LPSIFT   Layer parallel SIFT
LQP   Local Quaternary Patterns
LSc   Laplacian Sparse Coding
LSDA   Locality Sensitive Discriminant Analysis
MAE   Mean Absolute Error
MANOVA   Multivariate analysis of variance
MAPE   Mean Absolute Percent Error
MFA   Marginal Fisher Analysis
MI   Mutual Information
MSCNN   Multi-scale Convolutional Neural Networks
MSIFT   Multi-spectral SIFT
MTSVM   Multi-training Support Vector Machine
MVR   Majority Voting Rule
NIR   Near infrared
NN   Nearest Neighbor
NST   New Solar Telescope
OPPSIFT   Opponent SIFT
PCA   Principal Component Analysis
PHOG   Pyramid of Histograms of Orientation Gradients
PT   Power Transformation
QP   Quadratic Programming
RBF   Radial Basis Function
RBM   Restricted Boltzmann Machine
ReLU   Rectified Linear Unit
RF   Relevance Feedback
ROC   Receiver Operator Characteristic
RSVM   Reduced Support Vector Machine
SAC   Soft Assignment Coding

SD	Statistical Dependency
SDO	Solar Dynamic Observatory
SFFS	Sequential Floating Forward Selection
SFS	Sequential Forward Selection
SIFT	Scale-Invariant Feature Transform
SKML	Sparse Kernel Manifold Learner
SMO	Sequential Minimal Optimization
SPM	Spatial Pyramid Matching
SRC	Sparse Representation-based Classification
SRM	Structural Risk Minimization
SURF	Speeded Up Robust Features
SVM_{Active}	Support Vector Machine Active Learning
SVM	Support Vector Machines
TNVWP	Thresholding Normalized Visual Word Plausibility
t-SNE	t-Distributed Stochastic Neighbor Embedding
ULDA	Uncorrelated Linear Discriminant Analysis
VA	Video Analytics
VLAD	Vector of Locally Aggregated Descriptors
WLD	Weber Local Descriptors
WS-FV	Weber-SIFT Fisher Vector

Chapter 1
Feature Representation and Extraction for Image Search and Video Retrieval

Qingfeng Liu, Yukhe Lavinia, Abhishek Verma, Joyoung Lee, Lazar Spasovic and Chengjun Liu

Abstract The ever-increasing popularity of intelligent image search and video retrieval warrants a comprehensive study of the major feature representation and extraction methods often applied in image search and video retrieval. Towards that end, this chapter reviews some representative feature representation and extraction approaches, such as the Spatial Pyramid Matching (SPM), the soft assignment coding, the Fisher vector coding, the sparse coding and its variants, the Local Binary Pattern (LBP), the Feature Local Binary Patterns (FLBP), the Local Quaternary Patterns (LQP), the Feature Local Quaternary Patterns (FLQP), the Scale-invariant feature transform (SIFT), and the SIFT variants, which are broadly applied in intelligent image search and video retrieval.

1.1 Introduction

The effective methods in intelligent image search and video retrieval are often inter-disciplinary in nature, as they cut across the areas of probability, statistics, real analysis, digital signal processing, digital image processing, digital video processing, computer vision, pattern recognition, machine learning, and artificial intelligence,

Q. Liu (✉) · J. Lee · L. Spasovic · C. Liu (✉)
New Jersey Institute of Technology, Newark, NJ 07102, USA
e-mail: ql69@njit.edu

C. Liu
e-mail: chengjun.liu@njit.edu

J. Lee
e-mail: jo.y.lee@njit.edu

L. Spasovic
e-mail: spasovic@njit.edu

Y. Lavinia (✉) · A. Verma (✉)
California State University, Fullerton, CA 92834, USA
e-mail: ylavinia@csu.fullerton.edu

A. Verma
e-mail: averma@fullerton.edu

© Springer International Publishing AG 2017
C. Liu (ed.), *Recent Advances in Intelligent Image Search and Video Retrieval*,
Intelligent Systems Reference Library 121, DOI 10.1007/978-3-319-52081-0_1

1

Fig. 1.1 Example images from the Caltech-256 dataset

just to name a few. The applications of intelligent image search and video retrieval cover a broad range from web-based image search (e.g., photo search in Facebook) to Internet video retrieval (e.g., looking for a specific video in YouTube). Figure 1.1 shows some example images from the Caltech-256 dataset, which contains a set of 256 object categories with a total of 30,607 images [21]. Both the Caltech-101 and the Caltech-256 image datasets are commonly applied for evaluating the performance on image search and object recognition [21]. Figure 1.2 displays some video frames from the cameras installed along the highways. Actually, the New Jersey Department of Transportation (NJDOT) operates more than 400 traffic video cameras, but current traffic monitoring is mainly carried out by human operators. Automated traffic incident detection and monitoring is much needed as operator-based monitoring is often stressful and costly.

The ever-increasing popularity of intelligent image search and video retrieval thus warrants a comprehensive study of the major feature representation and extraction methods often applied in image search and video retrieval. Towards that end, this chapter reviews some representative feature representation and extraction approaches, such as the Spatial Pyramid Matching (SPM) [27], the soft assignment coding or kernel codebook [17, 18], the Fisher vector coding [24, 42], the sparse coding [53], the Local Binary Pattern (LBP) [40], the Feature Local Binary Patterns

Fig. 1.2 Example video frames from the cameras installed along the highways

(FLBP) [23, 31], the Local Quaternary Patterns (LQP) [22], the Feature Local Quaternary Patterns (FLQP) [22, 31], the Scale-invariant feature transform (SIFT) [35], and the SIFT variants, which are broadly applied in intelligent image search and video retrieval.

1.2   Spatial Pyramid Matching, Soft Assignment Coding, Fisher Vector Coding, and Sparse Coding

1.2.1   Spatial Pyramid Matching

The bag of visual words [13, 27] method starts with the k-means algorithm for deriving the dictionary and the hard assignment coding method for feature coding. One representative method is the spatial pyramid matching (SPM) [27] method, which enhances the discriminative capability of the conventional bag of visual words method by incorporating the spatial information.

Specifically, given the local feature descriptors $\mathbf{x}_i \in \mathbb{R}^n$, $(i = 1, 2, \ldots, m)$ and the dictionary of visual words $\mathbf{D} = [\mathbf{d}_1, \mathbf{d}_2, \ldots, \mathbf{d}_k] \in \mathbb{R}^{n \times k}$ derived from the k-means algorithm, the SPM method counts the frequency of the local features over the visual words and represents the image as a histogram using the following hard assignment coding method:

$$c_{ij} = \begin{cases} 1 & \text{if } j = \arg\min ||\mathbf{x}_i - \mathbf{d}_j||^2 \\ 0 & \text{otherwise} \end{cases} \tag{1.1}$$

In other words, the SPM method activates only one non-zero coding coefficient, which corresponds to the nearest visual word in the dictionary $\mathbf{D}$ for each local feature descriptor $\mathbf{x}_i$. And given one image I with T local feature descriptors, the corresponding image representation is the probability density estimation of all the local features $\mathbf{x}_i$ in this image I over all the visual words $\mathbf{d}_j$ based on the histogram of visual word frequencies as follows:

$$h = \left[\frac{1}{T} \sum_{i=1}^{m} c_{i1}, \frac{1}{T} \sum_{i=1}^{m} c_{i2}, \ldots, \frac{1}{T} \sum_{i=1}^{m} c_{ik} \right] \tag{1.2}$$

1.2.2 Soft Assignment Coding

The histogram estimation of the density function for the local features $\mathbf{x}_i$ over the visual words $\mathbf{d}_j$, which violates the ambiguous nature of local features, is a very coarse estimation. Therefore, the soft assignment coding [17, 18], or kernel codebook, is proposed as a more robust alternative to histogram.

Specifically, the soft-assignment coding of $\mathbf{c}_{ij}$ is defined as follows:

$$c_{ij} = \frac{exp(-||\mathbf{x}_i - \mathbf{d}_j||^2/2\sigma^2)}{\sum_{j=1}^{k} exp(-||\mathbf{x}_i - \mathbf{d}_j||^2/2\sigma^2)} \tag{1.3}$$

where σ is the smoothing parameter that controls the degree of smoothness of the assignment and $exp(\cdot)$ is the exponential function.

Consequently, given one image I with T local feature descriptors, the corresponding image representation is the probability density estimation of the all the local features $\mathbf{x}_i$ in this image I over all the visual words $\mathbf{d}_j$ based on the kernel density estimation using the Gaussian kernel $K(\mathbf{x}) = \frac{1}{\sqrt{2\pi\sigma}} exp(-\frac{\mathbf{x}^2}{2\sigma^2})$ as follows:

$$\begin{aligned} h &= \left[\frac{1}{T} \sum_{i=1}^{m} c_{i1}, \ldots, \frac{1}{T} \sum_{i=1}^{m} c_{ik} \right] \\ &= \left[\frac{1}{T} \sum_{i=1}^{T} w_i K(\mathbf{d}_1 - \mathbf{x}_i), \ldots, \frac{1}{T} \sum_{i=1}^{T} w_i K(\mathbf{d}_k - \mathbf{x}_i) \right] \end{aligned} \tag{1.4}$$

where σ in the Gaussian kernel plays the role of bandwidth in kernel density estimation, and $w_i = \frac{1}{\sum_{j=1}^{k} K(\mathbf{x}_i - \mathbf{d}_j)}$.

1.2.3 Fisher Vector Coding

The kernel density estimation in soft assignment coding is still error-prone due to its limitation in probability in density estimation. Recently, the Fisher vector method [24, 42] is proposed that the generative probability function of the local feature descriptors is estimated using a more refined model, namely the Gaussian mixture model (GMM). Then the GMM is applied to derive the Fisher kernel, which is incorporated into the kernel based support vector machine for classification. Fisher vector coding method [42] is essentially an explicit decomposition of the Fisher kernel.

As for dictionary learning, unlike the spatial pyramid matching and the soft assignment coding method, the Fisher vector coding method replaces the k-means algorithm with a Gaussian mixture model as follows:

$$\mu_\lambda(x) = \sum_{j=1}^{k} w_j g_j(x : \mu_j, \sigma_j) \tag{1.5}$$

where the parameter set $\lambda = \{w_j, \mu_j, \sigma_j, j = 1, 2, \ldots, k\}$ represents the mixture weight, the mean vector, and the covariance matrix of the Gaussian components, respectively. As a result, the visual words are no longer the centroids of the clusters, but rather the GMM components.

As for feature coding, the Fisher vector coding method applies the gradient score of the j−th component of the GMM over its parameters (μ_j is used here), instead of the hard/soft assignment coding methods as follows:

$$c_{ij} = \frac{1}{\sqrt{w_j}} \gamma_i(j) \sigma_j^{-1}(x_i - \mu_j) \tag{1.6}$$

where $\gamma_i(j) = \frac{w_j g_j(x_i)}{\sum_{t=1}^{k} w_t g_t(x_i)}$. As a result, given one image I with T local feature descriptors, the corresponding image representation, namely the Fisher vector, is the histogram of the gradient score of all the local features $\mathbf{x}_i$ in this image I:

$$
\begin{aligned}
h &= \left[\frac{1}{T} \sum_{i=1}^{m} c_{i1}, \ldots, \frac{1}{T} \sum_{i=1}^{m} c_{ik} \right] \\
&= \left[\frac{1}{T} \sum_{i=1}^{T} \frac{1}{\sqrt{w_1}} \gamma_i(1) \sigma_1^{-1}(x_i - \mu_1), \ldots \frac{1}{T} \sum_{i=1}^{T} \frac{1}{\sqrt{w_k}} \gamma_i(k) \sigma_k^{-1}(x_i - \mu_k) \right]
\end{aligned} \tag{1.7}
$$

1.2.4 Sparse Coding

The sparse coding method deals with the dictionary learning and the feature coding from the reconstruction point of view. Yang et al. [53] applied the sparse coding to learn a dictionary and a vector of coefficients for the feature coding.

Specifically, the sparse coding method is the optimization of the following objective function:

$$\min_{\mathbf{D,W}} \sum_{i=1}^{m} ||\mathbf{x}_i - \mathbf{Dw}_i||^2 + \lambda||\mathbf{w}_i||_1 \tag{1.8}$$

$$s.t. \quad ||\mathbf{d}_j|| \leq 1, (j = 1, 2, \ldots, k)$$

The sparse coding method applies a reconstruction criterion so that the original local feature descriptor can be reconstructed as a linear combination of the visual words in the dictionary and most of the coefficients are zero. Many methods are proposed for optimizing the objective function, such as the fast iterative shrinkage-thresholding algorithms (FISTA) [8], the efficient learning method [28], as well as the online learning method [36]. After the optimization, both the dictionary and the sparse coding are obtained. Then following the notation in the above sections, the sparse coding method derives the following coding:

$$c_{ij} = w_{ij} \tag{1.9}$$

where w_{ij} is an element in the sparse coding vector $\mathbf{w}_i$.

Consequently, given one image I with T local feature descriptors, the corresponding image representation is computed either using the average pooling method (h_{avg}) or the max pooling method (h_{max}) [46, 53] as follows:

$$h_{avg} = \left[\frac{1}{T} \sum_{i=1}^{m} c_{i1}, \ldots, \frac{1}{T} \sum_{i=1}^{m} c_{ik} \right] \tag{1.10}$$

$$h_{max} = [\max\{c_{i1}\}, \ldots, \max\{c_{ik}\}]$$

We have reviewed in the formal or mathematical means some representative feature representation and extraction methods for intelligent image search and video retrieval. Figure 1.3 shows in a more intuitive and graphical manner the comparison among the Spatial Pyramid Matching (SPM), the Soft Assignment Coding (SAC), the Fisher vector coding, and the sparse coding methods.

1.2.5 Some Sparse Coding Variants

There are a lot of variants of sparse coding methods that are proposed for addressing various issues in sparse coding. Wang et al. [46] proposed the locality-constrained linear coding (LLC) method that incorporates the local information in the feature coding process. Specifically, the LLC method incorporates a locality criterion into the sparse coding criterion to derive a new coding for the local feature descriptor that takes the local information into account. Gao et al. [16] further proposed the Laplacian sparse coding (LSC) that preserves both the similarity and the locality information among the local features. Specifically, the proposed LSC introduces a graph sparse coding criterion into the sparse coding criterion to derive a new coding

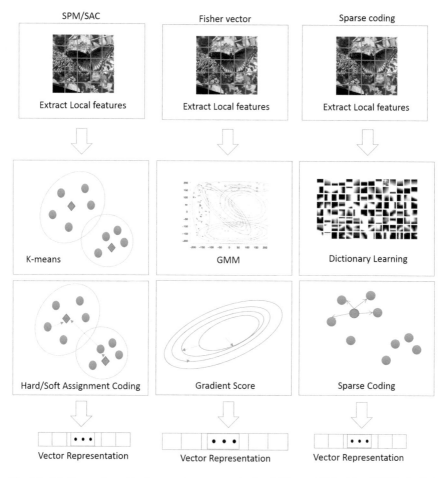

Fig. 1.3 Intuitive and graphical comparison among the Spatial Pyramid Matching (SPM), the Soft Assignment Coding (SAC), the Fisher vector coding, and the sparse coding methods

method to utilize the underlying structure of sparse coding. Zhou et al. [56] proposed a super vector coding method which takes advantage of the probability kernel method for sparse coding. Bo et al. [9] proposed a hierarchical sparse coding methods for image classification, which harnesses the hierarchical structure of the sparse coding.

In addition, many papers on sparse coding focus on developing efficient learning algorithms to derive the sparse coding and the dictionary [8, 20, 28, 36, 47–50], or exploring the data manifold structures [16, 46, 55]. For efficiency optimization, recent research applies screening rules for improving the computational efficiency of the sparse coding (lasso) problem. Ghaoui et al. [20] presented the SAFE screening rule for the lasso problem and sparse support vector machine. Xiang et al. [49] derived two new screening tests for large scale dictionaries and later proposed the DOME test [50] for the lasso problem. Wang et al. [48] proposed the Dual Polytope Projection

(DPP) for the lasso screening problem and later [47] proposed the "Slores" rule for sparse logistic regression screening.

1.3 Local Binary Patterns (LBP), Feature LBP (FLBP), Local Quaternary Patterns (LQP), and Feature LQP (FLQP)

The Local Binary Patterns (LBP) method, which uses the center pixel as a threshold and compares the pixels in its local neighborhood with the threshold to derive a gray-scale invariant texture description, has broad applications in feature representation and extraction for intelligent image search and video retrieval [38–40]. Specifically, some researchers apply the LBP method for facial image representation and then utilize the LBP texture features as a descriptor for face recognition [1, 2]. Other researchers propose a method that fuses the local LBP features, the global frequency features, and the color features for improving face recognition performance [34]. Yet others present new color LBP descriptors for scene and image texture classification [6].

The LBP method is popular for feature representation and extraction because of its computational simplicity and robustness to illumination changes. The limitation of the LBP method comes from the fact that it considers only its local pixels but not the features that are more broadly defined, among which are the edge pixels, the intensity peaks or valleys of an image, the color features [5, 30, 33, 43], and the wavelet features [11, 29, 32], just to name a few.

The Feature Local Binary Patterns (FLBP) method improves upon the LBP method by introducing features that complement the conventional neighborhood used to derive the LBP representation [23]. The FLBP method first defines a True Center (TC) and a Virtual Center (VC): The TC is the center pixel of a neighborhood and the VC is specified on the distance vector by a VC parameter and is used to replace the center pixel of the neighborhood [23]. The FLBP representation is then defined by comparing the pixels in the neighborhood of the true center with the virtual center [23]. It is shown that the LBP method is a special case of the FLBP method [23]: when the TC and VC parameters are zero, the FLBP method degenerates to the LBP method. There are two special cases of the FLBP method: when the VC parameter is zero, the method is called the FLBP1 method; and when the TC parameter is zero, the method is called the FLBP2 method [23]. As these FLBP methods encode both the local information and the features that are broadly defined, they are expected to perform better than the LBP method for feature representation and extraction for intelligent image search and video retrieval [23, 31].

The Local Quaternary Patterns (LQP) method augments the LBP method by encoding four relationships of the local texture [22]:

$$S_{lqp}(g_i, g_c, r) = \begin{cases} 11, \text{ if } g_i \geq g_c + r \\ 10, \text{ if } g_c \leq g_i < g_c + r \\ 01, \text{ if } g_c - r \leq g_i < g_c \\ 00, \text{ if } g_i < g_c - r \end{cases} \qquad (1.11)$$

where g_i and g_c represent the grey level of a neighbor pixel and the central pixel, respectively. $r = c + \tau g_c$ defines the radius of the interval around the central pixel and c is a constant and τ is a parameter to control the contribution of g_c to r. For efficiency, the LQP representation can be split into two binary codes, namely the upper half of LQP (ULQP) and the lower half of LQP (LLQP) [22]. As a result, the LQP method encodes more information of the local texture than the LBP method and the Local Ternary Patterns (LTP) method [22]. The Feature Local Quaternary Patterns (FLQP) method further encodes features that are broadly defined. Thus the FLQP method should further improve image search and video retrieval performance when used for feature representation and extraction [22].

1.4 Scale Invariant Feature Transform (SIFT) and SIFT Variants

SIFT [35] is one of the most commonly used local descriptors in intelligent image search and video retrieval. Its power lies in its robustness to affine distortion, viewpoints, clutters, and illumination changes. This makes SIFT an invaluable method in various computer vision applications such as face, object recognition, robotics, human activity recognition, panorama stitching, augmented reality, and medical image analysis.

The SIFT algorithm comprises the following steps: (1) scale space extrema detection, (2) keypoint localization, (3) orientation assignment, (4) keypoint descriptor construction, and (5) keypoint matching. To detect the peak keypoints, SIFT uses Laplacian of Gaussian (LoG), which acts as a space filter by detecting blobs in various scales and sizes. Due to its expensive computation, SIFT approximates LoG with Difference of Gaussian (DoG). The DoG is produced by computing the difference of Gaussian blurring on an image with two different scales that are represented in different octaves in the Gaussian pyramid. Following this, the keypoint extrema candidates are located by selecting each pixel in an image and comparing it with its 8 neighbors and the 9 pixels of its previous and next scales, amounting to 26 pixels to compare. Next is keypoint localization, which analyzes the keypoint candidates produced in the previous step. SIFT uses the Taylor series expansion to exclude the keypoints with low contrast, thus leaving the ones with strong interest points to continue on the next step. Orientation assignment is geared to achieve rotation invariance. At each keypoint, the central derivatives, the gradient magnitude, and the direction are computed, producing a weighted orientation histogram with 36 bins around the keypoint neighborhood. The most dominant, that is, the highest peak of the histograms, of the orientations is selected as the direction of the keypoint. To construct the keypoint descriptor, a 16×16 neighborhood region around a keypoint is selected to compute the keypoint relative orientation and magnitude. It is further divided into 16 subblocks, each with 4×4 size. An 8-bin orientation histogram is created for each subblock. The 16 histograms are concatenated to form a 128-dimension descriptor. Finally, keypoint matching is done by computing the Euclidean distance between two keypoints. First, a database of keypoints is constructed from the training images. Next, when a keypoint is to be matched, it is compared with the ones stored in a database. The Euclidean distance of these keypoints is computed and the database keypoint with minimum Euclidean distance is selected as the match.

Since its creation, many works have been dedicated to improve SIFT. The modifications are done in various steps of the SIFT algorithm and are proven to increase not only recognition rate but also speed. The following provides brief descriptions of the SIFT and SIFT like descriptors: Color SIFT, SURF, MSIFT, DSP-SIFT, LPSIFT, FAIR-SURF, Laplacian SIFT, Edge-SIFT, CSIFT, RootSIFT, and PCA-SIFT.

1.4.1 Color SIFT

The various color spaces such as RGB, HSV, rgb, oRGB, and YCbCr can be used to enhance SIFT performance [45]. The color SIFT descriptors are constructed by computing the 128 dimensional vector of the SIFT descriptor on the three channels, yielding 384 dimensional descriptors of RGB-SIFT, HSV-SIFT, rgb-SIFT, and YCbCr-SIFT.

Concatenating the three image components of the oRGB color space produces the oRGB-SIFT. Fusing RGB-SIFT, HSV-SIFT, rgb-SIFT, oRGB-SIFT, and YCbCr-SIFT generates the Color SIFT Fusion (CSF). Further fusion of the CSF and the grayscale SIFT produces the Color Grayscale SIFT Fusion (CGSF) [44, 45].

Results of the experiments on several grand challenge image datasets show that oRGB-SIFT descriptor improves recognition performance upon other color SIFT descriptors, the CSF, the CGSF, and the CGSF + PHOG descriptors perform better than the other color SIFT descriptors. The fusion of both Color SIFT descriptors (CSF) and Color Grayscale SIFT descriptor (CGSF) show significant improvement in the classification performance, which indicates that various color-SIFT descriptors and grayscale-SIFT descriptor are not redundant for image classification [44, 45].

1.4.2 SURF

Speeded Up Robust Features (SURF) [7] is a SIFT-based local feature detector and descriptor. SURF differs from SIFT in the following aspects. First, in scale space analysis, instead of using DoG, SURF uses a *fast Hessian detector* that is based on the Hessian matrix and implemented using box filters and integral images. Second, in orientation assignment, SURF computes Haar wavelet responses in the vertical and horizontal directions within the scale s and radius $6s$ from the interest points. Estimation of the dominant orientation is computed by summing the responses obtained through a sliding 60° angled window. Third, in extracting the descriptor, SURF forms a square region centered at an interest point and oriented according to the dominant orientation. The region is divided into sub-regions. For each sub-region, SURF then computes the sum of the Haar wavelet responses in the vertical and horizontal directions of selected interest points, producing a 64-dimensional SURF feature descriptor.

To improve its discriminative power, the SURF descriptor can be extended to 128 dimensions. This is done by separating the summation computation for d_x and $|d_x|$ according to the sign of d_y ($d_y < 0$ and $d_y \geq 0$) and the computation for d_y and $|d_y|$ according to the sign of d_x. The result is a doubled number of features, creating a descriptor with increased discriminative power. SURF-128, however, performs slightly slower than SURF-64, although still faster than SIFT. This compromise turns out to be advantageous with SURF-128 achieving higher recognition rate than SURF-64 and SIFT. Results in [7] show SURF's improved performance on standard image datasets as well as on imagery obtained in the context of real life object detection application.

1.4.3 MSIFT

Multi-spectral SIFT (MSIFT) [10] takes advantage of near infrared (NIR) to enhance recognition. The NIR occupies the 750–1100 nm region on the wavelength spectrum, and silicon, the primary semiconductor in digital camera chip, is known to have high sensitivity to this region.

The MSIFT descriptor is developed by first decorrelating the RGB-NIR 4-dimensional color vector, followed by linear transformation of the resulting decorrelated components. This produces four components with the first being achromatic (luminance) with roughly the same amount of R, G, B, and high NIR, and the other three consisting of various spectral difference of R, G, B, and NIR. Next, forming the multi-spectral keypoint is done through Gaussian extrema detection in the achromatic first component. It is then followed by creation of 4×4 histogram of gradient orientations of each channel. The color bands are normalized and concatenated to form the final descriptor. Since the resulting RGB-NIR descriptors dimensionality amounts to 512, a PCA dimensionality reduction is applied.

The immediate application of MSIFT is to solve scene recognition problems. As noted above, with silicon's high sensitivity to the NIR region, an MSIFT equipped digital camera would be able to offer enhanced intelligent scene recognition features to users.

1.4.4 DSP-SIFT

Domain Size Pooling (DSP) SIFT [14] defies the traditional scale space step in SIFT descriptor construction and replaces this step with size space. Instead of forming the descriptor from a single selected scaled lattice, in DSP-SIFT, multiple lattices with various domain sizes are sampled, as shown in Fig. 1.4a. To make them all in the same size, each lattice sample is rescaled, making these multiple lattice samples differ in scales, although uniform in size (Fig. 1.4b). Pooling of the gradient orientations is done across these various locations and scales (Fig. 1.4c). These gradient orientations are integrated and normalized by applying a uniform density function (Fig. 1.4d), which then yields the final DSP-SIFT descriptor (Fig. 1.4e).

Authors in [14] report that DSP-SIFT outperforms SIFT by a wide margin and furthermore it outperforms CNN by 21% on the Oxford image matching dataset and more than 5% on the Fischer dataset.

1.4.5 LPSIFT

The layer parallel SIFT (LPSIFT) [12] seeks to implement SIFT on real time devices by reducing the computational cost, time latency, and memory storage of the original SIFT. The modification is done primarily on the most expensive steps of the original SIFT algorithm: scale space Gaussian extrema detection and keypoint localization. The main cause of this bottleneck, as it is observed, is the data dependency of scale image computation. The Gaussian blurring computation on each new image is done sequentially and this proves to cause a large memory expense for the next step in the pipeline: computation of the Difference of Gaussian (DoG). As the DoG pyramid requires at least three scale images to complete, a candidate image needs to wait for the Gaussian blurring operation to compute the next two images, and thus the image must be stored in memory.

LPSIFT solves the problem by introducing layer parallel to Gaussian pyramid construction. To handle simultaneous computation on multiple images, the layer parallel simply merges the kernels on the same level and forwards them to the DoG

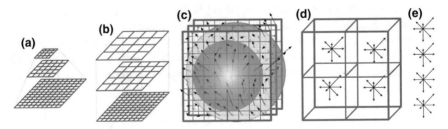

Fig. 1.4 DSP-SIFT methodology

operation. This trick significantly reduces the time latency caused by the sequential flow and also the memory cost needed to store the images. The merged kernel, however, can potentially expand to a size that would cause an increased computational cost. To avoid it, LPSIFT uses integral images that are implemented on modified box kernels of sizes 3×3, 5×5, and 7×7.

The next modification on the original SIFT algorithm takes place in the contrast test. The original SIFT algorithm contrast test aims to remove low contrast keypoints using the Taylor expansion, which is high in complexity. LPSIFT circumvents the use of the expensive Taylor series by modifying the algorithm to exclude low brightness instead of low contrast. The rationale is that low brightness is an accompanying characteristic of low contrast and thus excluding low brightness candidate keypoints would also exclude low contrast candidates.

With this method, LPSIFT manages to reduce 90% of computational cost and 95% of memory usage.

1.4.6 FAIR-SURF

Fully Affine InvaRiant SURF (FAIR-SURF) [41] modifies the SURF algorithm to be fully affine invariant by using techniques used in ASIFT [37]. Like ASIFT, FAIR-SURF uses two camera axis parameters simulation to generate images. It then applies the SURF algorithm for feature extraction. While ASIFT uses finite rotation and tilts to simulate camera angles in generating images and therefore runs much slower, FAIR-SURF selects only certain rotation and tilt angles, thus improving the speed.

The selection process is described as follows. It is observed that SURF maintains fully affine invariant until a certain low angle and that 2 is a balanced value between accuracy and sparsity [37]. The angles under which SURF is fully affine invariant are chosen so as to extend this fully affine invariant trait to the resulting FAIR-SURF. Thus, a list of rotation and tilt angles are formed.

To reduce the number of angles, Pang et al. [41] applied the modified SURF to extract features and perform image matching. They then compared the matching results with the list and selected several to be tested on other images. The matching results became the final list of rotation and tilt angles that are used to simulate images. These images are then used in the next steps of the SURF algorithm.

By using selected images that are simulated using angles under which SURF is fully affine invariant, FAIR-SURF achieves full affine invariance. Its keypoints produce higher matches compared to SURF and ASIFT and its runtime, although 1.3 times slower than the original SURF, is still faster than ASIFT.

1.4.7 Laplacian SIFT

In visual search on resource constrained devices, a descriptor's matching ability and compactness are critical. Laplacian SIFT [51] aims to improve these qualities by preserving the nearest neighbor relationship of the SIFT features. This is because it is observed that the nearest neighbors contain important information that can be used to improve matching ability. The technique utilizes graph embedding [52] and specifically the Laplacian embedding, to preserve more nearest neighbor information and reduce the dimensionality.

The image retrieval process using Laplacian SIFT is implemented in two separate segments: data preprocessing and query processing. Data preprocessing takes a set of feature points and employs Laplacian embedding to reduce the original SIFT 128 feature dimension to a desired feature dimension. The experiment uses a 32-bit representation with 4 dimensions and 8 bits quantization per dimension. Two kd-trees are then created on the resulting feature points. These trees are formed to discard the feature points located at the leaf node boundaries.

Query processing takes a set of query feature points and selects the features according to the feature selection algorithms. Two leaf nodes are located and merged. On each leaf node, the algorithm performs nearest neighbor matching and compares the results with a predetermined threshold value.

1.4.8 Edge-SIFT

Edge-SIFT [54] is developed to improve efficiency and compactness of large scale partial duplicate image search on mobile platforms. Its main idea is to use binary edge maps to suppress memory footprint. Edge-SIFT focuses on extracting edges since they preserve spatial cues necessary for identification and matching yet sparse enough to maintain compactness.

The first step to construct Edge-SIFT is creating image patches that are centered at interest points. These patches are normalized to make them scale and rotation invariant. Scale invariance is achieved through resizing each patch to a fixed size, while rotation invariance is through aligning the image patches to make their dominant orientations uniform. An edge extractor is then used to create the binary edge maps with edge pixel values of 1 or 0. The edge map is further decomposed into four sub-edge maps with four different orientations. To overcome sensitivity to registration errors in the vertical direction, an edge pixel expansion is applied in the vertical direction of the maps orientation.

The resulting initial Edge-SIFT is further compressed. Out of the most compact bins, the ones with the highest discriminative power are selected using RankBoost [15]. To reduce the speed of similarity computation, the results are stored in a lookup table. The final Edge-SIFT was proven to be more compact, efficient, and accurate than the original SIFT. The direct application of the algorithm includes landmark 3D construction and image panoramic view generator.

1.4.9 CSIFT

CSIFT, or Colored SIFT [3], uses color invariance [19] to build the descriptor. As the original SIFT is designed to be applied to gray images, modifying the algorithm to apply to color images is expected to improve the performance. The color invariance model used in CSIFT is derived from the Kubelka Munk theory of photometric reflectance [26] and describes color invariants under various imaging conditions and assumptions concerning illumination intensity, color, and direction, surface orientation, highlights, and viewpoint. The color invariants can be calculated from the RGB color space using the Gaussian color model to represent spatial spectral information [19].

CSIFT uses these color invariants to expand the input image space and invariance for keypoint detection. The gradient orientation is also computed from these color invariants. The resulting descriptor is robust to image translation, rotation, occlusion, scale, and photometric variations. Compared to the grayscale based original SIFT, CSIFT generates higher detection and matching rate.

1.4.10 RootSIFT

RootSIFT [4] is based on the observation that in comparing histograms, the Hellinger kernel generates superior results to the Euclidean in image categorization, object and texture classification. As the original SIFT uses the Euclidean distance to compare histograms, it is expected that using the Hellinger kernel would improve the results. The original SIFT descriptor can be converted to RootSIFT using the Hellinger kernel.

The Hellinger kernel implementation in RootSIFT requires only two additional steps after the original SIFT: (1) L1 normalization of the original SIFT vector, and (2) taking the square root of each element. By executing these steps, the resulting SIFT vectors are L2 normalized. Classification results by applying the RootSIFT on Oxford buildings dataset and the PASCAL VOC image dataset show improvement upon SIFT [4].

1.4.11 PCA-SIFT

PCA-SIFT [25] undergoes the same few steps of the SIFT algorithm (scale space extrema detection, keypoint localization, and gradient orientation assignment) but modifies the keypoint descriptor construction. PCA-SIFT uses a projection matrix to create the PCA-SIFT descriptor. To form this projection matrix, keypoints are selected and rotated towards their dominant orientation, and a 41×41 patch that is centered at each keypoint is created. The vertical and horizontal gradients of these patches are computed, forming an input vector of size $2 \times 39 \times 39$ with 3,042 elements for each patch. The covariance matrix of these vectors is computed, followed by the eigenvectors and eigenvalues of the matrices. The top n eigenvectors are then selected to construct $n \times 3,042$ projection matrix, with n being an empirically determined value. This projection matrix is computed once and stored.

The descriptor is formed by extracting a 41×41 patch around a keypoint, rotating it to its dominant orientation, creating a normalized 3,042 element gradient image vector from the horizontal and vertical gradients, and constructing a feature vector by multiplying the gradient image vector with the stored $n \times 3,042$ projection matrix. The resulting PCA-SIFT descriptor is of size n. With $n = 20$ as the feature space size, PCA-SIFT outperformed the original SIFT that uses 128-element vectors. Results show that using this descriptor in an image retrieval application results in increased accuracy and faster matching [25].

1.5   Conclusion

Some representative feature representation and extraction methods with broad applications in intelligent image search and video retrieval are reviewed. Specifically, the density estimation based methods encompass the Spatial Pyramid Matching (SPM) [27], the soft assignment coding or kernel codebook [17, 18], and the Fisher vector coding [24, 42]. The reconstruction based methods consist of the sparse coding [53] and the sparse coding variants. The local feature based methods inculde the Local Binary Pattern (LBP) [40], the Feature Local Binary Patterns (FLBP) [23, 31], the Local Quaternary Patterns (LQP) [22], and the Feature Local Quaternary Patterns (FLQP) [22, 31]. And finally the invariant methods contain the Scale-invariant feature transform (SIFT) [35], and the SIFT like descriptors, such as the Color SIFT, the SURF, the MSIFT, the DSP-SIFT, the LPSIFT, the FAIR-SURF, the Laplacian SIFT, the Edge-SIFT, the CSIFT, the RootSIFT, and the PCA-SIFT.

References

1. Ahonen, T., Hadid, A., Pietikainen, M.: Face recognition with local binary patterns. In: 8th European Conference on Computer Vision, Prague, Czech Republic, pp. 469–481 (2004)
2. Ahonen, T., Hadid, A., Pietikainen, M.: Face description with local binary patterns: application to face recognition. IEEE Trans. Pattern Anal. Mach. Intell. **28**(12), 2037–2041 (2006)
3. Aly, A., Farag, A.: Csift: a sift descriptor with color invariant characteristics. In: IEEE International Conference on Computer Vision and Pattern Recognition, New York, NY, pp. 1978–1983 (2006)

4. Arandjelovic, R., Zisserman, A.: Three things everyone should know to improve object retrieval. In: IEEE International Conference on Computer Vision and Pattern Recognition, Providence, RI, pp. 2911–2918 (2012)
5. Banerji, S., Sinha, A., Liu, C.: New image descriptors based on color, texture, shape, and wavelets for object and scene image classification. Neurocomputing **117**, 173–185 (2013)
6. Banerji, S., Verma, A., Liu, C.: Novel color LBP descriptors for scene and image texture classification. In: 15th International Conference on Image Processing, Computer Vision, and Pattern Recognition, Las Vegas, Nevada, USA (2011)
7. Bay, H., Tuytelaars, T., Van Gool, L.V.: SURF: speeded up robust features. Comput. Vision Image Underst. **110**(3), 346–359 (2008)
8. Beck, A., Teboulle, M.: A fast iterative shrinkage-thresholding algorithm for linear inverse problems. SIAM J. Imaging Sci. **2**(1), 183–202 (2009)
9. Bo, L., Ren, X., Fox, D.: Hierarchical matching pursuit for image classification: architecture and fast algorithms. In: Advances in Neural Information Processing Systems, pp. 2115–2123 (2011)
10. Brown, M., Ssstrunk, S.: Multi-spectral sift for scene category recognition. In: IEEE International Conference on Computer Vision and Pattern Recognition, Providence, RI, pp. 177–184 (2011)
11. Chen, S., Liu, C.: Eye detection using discriminatory haar features and a new efficient svm. Image Vision Comput. **33**, 68–77 (2015)
12. Chiu, L., Chang, T.S., Chen, J.Y., Chang, N.Y.C.: Fast sift design for real-time visual feature extraction. IEEE Trans. Image Process. **22**(8), 3158–3167 (2013)
13. Csurka, G., Dance, C., Fan, L., Willamowski, J., Bray, C.: Visual categorization with bags of keypoints. In: Workshop on Statistical Learning in Computer Vision, Prague (2004)
14. Dong, J., Soatto, S.: Domain-size pooling in local descriptors: Dsp-sift. In: IEEE International Conference on Computer Vision and Pattern Recognition, Boston, MA (2015)
15. Freund, Y., Iyer, R., Schapire, R.E., Singer, Y.: An efficient boosting algorithm for combining preferences. J. Mach. Learn. Res. **4**, 933–969 (2003)
16. Gao, S., Tsang, I.W.H., Chia, L.T.: Laplacian sparse coding, hypergraph laplacian sparse coding, and applications. IEEE Trans. Pattern Anal. Mach. Intell. **35**(1), 92–104 (2013)
17. Gemert, J.C., Geusebroek, J.M., Veenman, C.J., Smeulders, A.W.: Kernel codebooks for scene categorization. In: ECCV, pp. 696–709 (2008)
18. van Gemert, J.C., Veenman, C.J., Smeulders, A.W.M., Geusebroek, J.M.: Visual word ambiguity. IEEE Trans. Pattern Anal. Mach. Intell. **32**(7), 1271–1283 (2010)
19. Geusebroek, J., Boomgaard, R.v.d., Smeulders, A., Geerts, H.: Color invariance. IEEE Trans. Pattern Anal. Mach. Intell. **23**(12), 1338–1350 (2001)
20. Ghaoui, L.E., Viallon, V., Rabbani, T.: Safe feature elimination in sparse supervised learning. Technical report UC/EECS-2010-126, EECS Department, University of California at Berkeley (2010)
21. Griffin, G., Holub, A., Perona, P.: Caltech-256 object category dataset. Technical report, California Institute of Technology (2007)
22. Gu, J., Liu, C.: Local quaternary patterns and feature local quaternary patterns. In: 16th International Conference on Image Processing, Computer Vision, and Pattern Recognition, Las Vegas, Nevada, USA (2012)
23. Gu, J., Liu, C.: Feature local binary patterns with application to eye detection. Neurocomputing **113**, 138–152 (2013)
24. Jegou, H., Perronnin, F., Douze, M., Sanchez, J., Perez, P., Schmid, C.: Aggregating local image descriptors into compact codes. IEEE Trans. Pattern Anal. Mach. Intell. **34**(9), 1704–1716 (2012)
25. Ke, Y., Sukthankar, R.: Pca-sift: a more distinctive representation for local image descriptors. In: IEEE International Conference on Computer Vision and Pattern Recognition, Washington, DC, vol. 2, pp. 506–513 (2004)
26. Kubelka, P.: New contribution to the optics of intensely light-scattering materials, part i. J. Opt. Soc. Am. **38**(5), 448–457 (1948)

27. Lazebnik, S., Schmid, C., Ponce, J.: Beyond bags of features: spatial pyramid matching for recognizing natural scene categories. In: CVPR, pp. 2169–2178 (2006)
28. Lee, H., Battle, A., Raina, R., Ng, A.Y.: Efficient sparse coding algorithms. In: NIPS, pp. 801–808 (2007)
29. Liu, C.: Gabor-based kernel PCA with fractional power polynomial models for face recognition. IEEE Trans. Pattern Anal. Mach. Intell. **26**(5), 572–581 (2004)
30. Liu, C.: Effective use of color information for large scale face verification. Neurocomputing **101**, 43–51 (2013)
31. Liu, C., Mago, V. (eds.): Cross Disciplinary Biometric Systems. Springer, Berlin (2012)
32. Liu, C., Wechsler, H.: Gabor feature based classification using the enhanced Fisher linear discriminant model for face recognition. IEEE Trans. Image Process. **11**(4), 467–476 (2002)
33. Liu, Q., Puthenputhussery, A., Liu, C.: A novel inheritable color space with application to kinship verification. In: the IEEE Winter Conference on Applications of Computer Vision, Lake Placid, New York (2016)
34. Liu, Z., Liu, C.: Fusion of color, local spatial and global frequency information for face recognition. Pattern Recognit. **43**(8), 2882–2890 (2010)
35. Lowe, D.G.: Distinctive image features from scale-invariant keypoints. Int. J. Comput. Vis. **60**(2), 91–110 (2004)
36. Mairal, J., Bach, F., Ponce, J., Sapiro, G.: Online dictionary learning for sparse coding. In: ICML, p. 87 (2009)
37. Morel, J., Yu, G.: Asift: a new framework for fully affine invariant image comparison. SIAM J. Imaging Sci. **2**(2), 438–469 (2009)
38. Ojala, T., Pietikainen, M., Harwood, D.: Performance evaluation of texture measures with classification based on kullback discrimination of distributions. In: Proceedings of the 12th IAPR International Conference on Pattern Recognition, Jerusalem, Israel, pp. 582–585 (1994)
39. Ojala, T., Pietikainen, M., Harwood, D.: A comparative study of texture measures with classification based on feature distributions. Pattern Recognit. **29**(1), 51–59 (1996)
40. Ojala, T., Pietikainen, M., Maenpaa, T.: Multiresolution gray-scale and rotation invariant texture classification with local binary patterns. IEEE Trans. Pattern Anal. Mach. Intell. **24**(7), 971–987 (2002)
41. Pang, Y., Lia, W., Yuanb, Y., Panc, J.: Fully affine invariant surf for image matching. Neurocomputing **85**(8), 6–10 (2012)
42. Sanchez, J., Perronnin, F., Mensink, T., Verbeek, J.J.: Image classification with the fisher vector: theory and practice. Int. J. Comput. Vis. **105**(3), 222–245 (2013)
43. Sinha, A., Banerji, S., Liu, C.: New color GPHOG descriptors for object and scene image classification. Mach. Vis. Appl. **25**(2), 361–375 (2014)
44. Verma, A., Liu, C.: Novel EFM- KNN classifier and a new color descriptor for image classification. In: 20th IEEE Wireless and Optical Communications Conference (Multimedia Services and Applications), Newark, New Jersey, USA (2011)
45. Verma, A., Liu, C., Jia, J.: New color SIFT descriptors for image classification with applications to biometrics. Int. J. Biom. **1**(3), 56–75 (2011)
46. Wang, J., Yang, J., Yu, K., Lv, F., Huang, T.S., Gong, Y.: Locality-constrained linear coding for image classification. In: CVPR, pp. 3360–3367 (2010)
47. Wang, J., Zhou, J., Liu, J., Wonka, P., Ye, J.: A safe screening rule for sparse logistic regression. In: Advances in Neural Information Processing Systems, pp. 1053–1061 (2014)
48. Wang, J., Zhou, J., Wonka, P., Ye, J.: Lasso screening rules via dual polytope projection. In: Advances in Neural Information Processing Systems, pp. 1070–1078 (2013)
49. Xiang, Z., Xu, H., Ramadge, P.: Learning sparse representations of high dimensional data on large scale dictionaries. Adv. Neural Inf. Process. Syst. **24**, 900–908 (2011)
50. Xiang, Z.J., Ramadge, P.J.: Fast lasso screening tests based on correlations. In: 2012 IEEE International Conference on Acoustics, Speech and Signal Processing, ICASSP 2012, pp. 2137–2140 (2012)
51. Xin, X., Li, Z., Katsaggelos, A.: Laplacian sift in visual search. In: IEEE International Conference on Acoustics, Speech, and Signal Processing, Kyoto, Japan (2012)

52. Yan, S., Xu, D., Zhang, B., Zhang, H.J., Yang, Q., Lin, S.: Graph embedding and extensions: a general framework for dimensionality reduction. IEEE Trans. Pattern Anal. Mach. Intell. (2007)
53. Yang, J., Yu, K., Gong, Y., Huang, T.S.: Linear spatial pyramid matching using sparse coding for image classification. In: CVPR, pp. 1794–1801 (2009)
54. Zhang, S., Tian, Q., Lu, K., Huang, Q., Gao, W.: Edge-sift: discriminative binary descriptor for scalable partial-duplicate mobile search. IEEE Trans. Image Process. 29(1), 40–51 (2013)
55. Zheng, M., Bu, J., Chen, C., Wang, C., Zhang, L., Qiu, G., Cai, D.: Graph regularized sparse coding for image representation. IEEE Trans. Image Process. 20(5), 1327–1336 (2011)
56. Zhou, X., Yu, K., Zhang, T., Huang, T.S.: Image classification using super-vector coding of local image descriptors. In: Proceedings of the 11th European Conference on Computer Vision: Part V, pp. 141–154 (2010)

Chapter 2
Learning and Recognition Methods for Image Search and Video Retrieval

Ajit Puthenputhussery, Shuo Chen, Joyoung Lee, Lazar Spasovic and Chengjun Liu

Abstract Effective learning and recognition methods play an important role in intelligent image search and video retrieval. This chapter therefore reviews some popular learning and recognition methods that are broadly applied for image search and video retrieval. First some popular deep learning methods are discussed, such as the feedforward deep neural networks, the deep autoencoders, the convolutional neural networks, and the Deep Boltzmann Machine (DBM). Second, Support Vector Machine (SVM), which is one of the popular machine learning methods, is reviewed. In particular, the linear support vector machine, the soft-margin support vector machine, the non-linear support vector machine, the simplified support vector machine, the efficient Support Vector Machine (eSVM), and the applications of SVM to image search and video retrieval are discussed. Finally, other popular kernel methods and new similarity measures are briefly reviewed.

2.1 Introduction

The applications in intelligent image search and video retrieval cover all corners of our society from searching the web to scientific discovery and societal security. For example, the New Solar Telescope (NST) at the Big Bear Solar Observatory (BBSO) produces over one terabytes of data daily, and the Solar Dynamic Observatory (SDO)

A. Puthenputhussery (✉) · J. Lee · L. Spasovic · C. Liu (✉)
New Jersey Institute of Technology, Newark, NJ 07102, USA
e-mail: avp38@njit.edu

C. Liu
e-mail: chengjun.liu@njit.edu

J. Lee
e-mail: jo.y.lee@njit.edu

L. Spasovic
e-mail: spasovic@njit.edu

S. Chen (✉)
The Neat Company, Philadelphia, PA 19103, USA
e-mail: sc77@njit.edu

© Springer International Publishing AG 2017
C. Liu (ed.), *Recent Advances in Intelligent Image Search and Video Retrieval*,
Intelligent Systems Reference Library 121, DOI 10.1007/978-3-319-52081-0_2

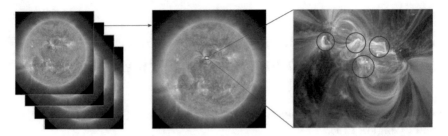

Fig. 2.1 Video analysis in solar physics for solar event detection

launched by NASA generates about four Terabytes of video data per day. Such huge amount of video data requires new digital image and video analysis techniques to advance the state-of-the-art in solar science. Motion analysis methods based on the motion field estimation and Kalman filtering may help characterize the dynamic properties of solar activities in high resolution. Figure 2.1 shows an example for solar features and events detection. First, a sequence of SDO images is processed based on motion field analysis to locate candidate regions of interest for features and events. The differential techniques that apply the optical flow estimation and the feature-based techniques that utilize feature matching and Kalman tracking approaches may be applied to locate the candidate regions for features and events. Other techniques, such as the conditional density propagation method and the particle filtering method, will further enhance the localization performance. Second, a host of feature and event detection and recognition methods will then analyze the candidate regions for detecting features and events as indicated in the right image in Fig. 2.1.

Another example of video-based applications for societal security is police body-worn cameras, which present an important and innovative area of criminal justice research with the potential to significantly advance criminal justice practice. The Community Policing Initiative proposed by the White House would provide a 50% match to states that purchase such cameras [83], and the recent Computing Community Consortium whitepaper on body-worn cameras released by a panel of computer vision experts and law enforcement personnel recommends increased research funding for technology development [17]. Figure 2.2 shows the idea of innovative police body-worn cameras that recognize their environment. Specifically, advanced face detection and facial recognition technologies, which are robust to challenging factors such as variable illumination conditions, may be applied for suspect detection in video to improve public safety and well-being of communities. The state-of-the-art image indexing and video retrieval methods should be utilized for searching, indexing, and triaging the large amount of video data in order to meet various criminal justice needs, such as forensic capabilities and the freedom of information act (FOIA) services.

This chapter reviews some representative learning and recognition methods that have broad applications in intelligent image search and video retrieval. We first discuss some popular deep learning methods, such as the feedforward deep neural

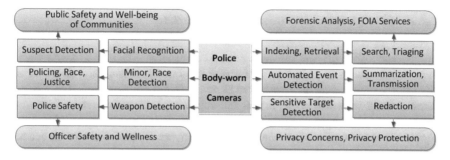

Fig. 2.2 Innovative police body-worn cameras that recognize their environment

networks [35, 49], the deep autoencoders [2, 81], the convolutional neural networks [23], and the Deep Boltzmann Machine (DBM) [22, 61]. We then discuss one of the popular machine learning methods, namely, Support Vector Machine (SVM) [77]. Specifically we review the linear support vector machine [78], the soft-margin support vector machine [78], the non-linear support vector machine [77], the simplified support vector machine [11, 54], the efficient Support Vector Machine (eSVM) [13, 14], and the applications of SVM to image search and video retrieval [25, 37, 52, 62, 75, 87]. We finally briefly review some other popular kernel methods and new similarity measures [40, 41, 44].

2.2 Deep Learning Networks and Models

Deep artificial neural network is an emerging research area in computer vision and machine learning and has gained increasing attention in recent years. With the advent of big data, the need for efficient learning methods to process an enormous number of images and videos for different kinds of visual applications has greatly increased. The task of image classification is a fundamental and important computer vision and machine learning problem wherein after learning a model based on a set of training and validation data, the labels for the test data have to be predicted. Image classification is a challenging problem as there can be many variations in the background noise, illumination conditions, as well as multiple poses, distortions and occlusions in the image. In recent years, different deep learning methods such as the feedforward deep neural networks [35, 49], the deep autoencoders [2, 81], the convolutional neural networks [23], and the Deep Boltzmann Machine (DBM) [22, 61], have been shown to achieve good performance for image classification problems. One possible reason for the feasibility of the deep learning methods is due to the discovery of multiple levels of representation within an image that leads to a better understanding of the semantics of the image.

2.2.1 Feedforward Deep Neural Networks

In this section, we discuss the architecture, the different layers, and some widely used activation functions in the feedforward deep neural networks. A feedforward deep neural network can be considered as an ensemble of many units that acts as a parametric function with many inputs and outputs to learn important features from the input image. Feedforward deep neural networks are also called multilayer perceptrons [59] and typically contain many hidden layers. Figure 2.3 shows the general architecture of a deep feedforward neural network with N hidden layers. Now let's consider a feedforward network with one hidden layer. Let the input layer be denoted as $\mathbf{I}$, the output layer as $\mathbf{O}$, the hidden layer as $\mathbf{H}$, and the weight vector for connections from the input layer to the hidden layer as $\mathbf{W}_1$. Therefore, the hidden unit vector can be computed as $\mathbf{H} = f(\mathbf{W}_1^T\mathbf{I} + \mathbf{b}_1)$, where $\mathbf{b}_1$ is the bias vector for the hidden layer and $f(.)$ is the activation function. Similarly, the output vector can be computed as $\mathbf{O} = f(\mathbf{W}_2^T\mathbf{H} + b_2)$, where $\mathbf{W}_2$ is the weight matrix for connections from the hidden layer to the output layer and $\mathbf{b}_2$ is the bias vector for the output layer.

The feedforward deep neural network can be optimized with many different optimization procedures but a common approach is to use momentum based stochastic gradient descent. An activation function takes in an input and performs some mathematical operation so that the output lies within some desired range. We next discuss some activation functions that are commonly used in the literature for deep neural networks. The rectified linear unit (ReLU) is the most popular activation function used for deep neural networks. It acts as a function that thresholds the input value at zero and has the mathematical form $g(y) = \max(0, y)$. Some advantages of the

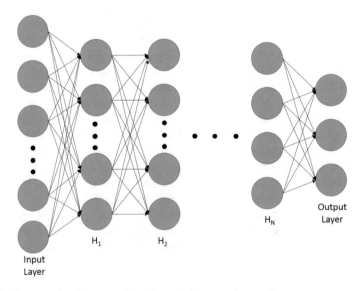

Fig. 2.3 The general architecture of feedforward deep neural networks

ReLU include that it helps the stochastic gradient descent process to converge faster [35] and can be implemented as a lightweight operation. Many different variants of the ReLU have been proposed to improve upon the ReLU. A leaky ReLU was propose by Maas et al. [49] wherein the function would never become zero but will be equal to a small constant. The leaky ReLU has the form $f(y) = \max(0, y) + c \min(0, y)$, where c is a small constant. Another variant known as the PReLU was proposed by He et al. [27] which considers c as a parameter learned during the training process. The sigmoid activation function produces an output in the range between 0 and 1 and has the form $\sigma(y) = 1/(1 + e^{-y})$. Some issues with the sigmoid activation function are that the output produced is not centered and it reduces the gradient to zero. The tanh activation function has the mathematical form $\tanh(y) = 2\sigma(2y) - 1$ and is a scaled version of the sigmoid activation function. It produces a centered output between -1 and 1, and overcomes the disadvantage of the sigmoid activation function.

2.2.2 Deep Autoencoders

The autoencoders are based on an unsupervised learning algorithm to develop an output representation that is similar to the input representation with the objective of minimizing the loss of information. Figure 2.4 shows the general architecture of a deep autoencoder where encoding takes place to transform the input vector into a compressed representation and decoding tries to reconstruct the original representation with minimum distortion [2, 81]. Let I be the set of training vectors, and the autoencoder problem may be formulated as follows: finding $\{\mathbf{W}_1, \mathbf{W}_2, b_1, b_2\}$ from $\mathbf{H} = f(\mathbf{W}_1^T \mathbf{I} + \mathbf{b}_1)$ and $\mathbf{O} = f(\mathbf{W}_2^T \mathbf{H} + b_2)$, such that $\mathbf{O}$ and $\mathbf{I}$ are similar with the

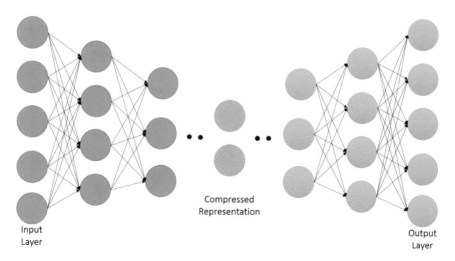

Input
Layer

Compressed
Representation

Output
Layer

Fig. 2.4 The general architecture of a deep autoencoder

minimum loss of information. A popular optimization procedure used for solving the autoencoders is the back propagation method for computing the gradient weights.

Different variants of autoencoders have been proposed so as to make them suitable for different applications. A sparse deep autoencoder [21, 57] integrates a sparsity criterion into the objective function to learn feature representation from images. Another variant is a denoising autoencoder [5] that is trained to reconstruct the correct output representation from a corrupted input data point. Deep autoencoders have been extensively applied for dimensionality reduction and manifold learning with applications to different visual classification tasks [21, 57].

2.2.3 Convolutional Neural Networks (CNNs)

In this section, we discuss the different layers, the formation of different layer blocks in a Convolutional Neural Network (CNN) [23] and some state-of-the-art CNNs [28, 35, 68, 70, 88] for the ImageNet challenge. A CNN is similar to a regular neural network but is more specifically designed for images as input and uses a convolution operation instead of a matrix multiplication. The most common layers in a CNN are the input layer, the convolution layer, the rectified linear unit (ReLU) layer, the pooling layer, and a fully connected layer. The convolution layer computes the dot product between the weights and a small region connected to the input image, whereas the ReLU layer performs the elementwise activation using the function $f(y) = \max(0, y)$. The pooling layer is used to reduce the spatial dimensions as we go deeper into the CNN, and finally the fully connected layer computes the class score for every class label of the dataset. An example of a convolutional neural network architecture is shown in Fig. 2.5 that contains two convolutional and pooling layers followed by an ReLU layer and a fully connected layer.

The ImageNet Large Scale Visual Recognition Challenge (ILSVRC) [60] is a challenging and popular image database having more than a million train images and 100,000 test images. Table 2.1 shows the performance of different CNNs on the ImageNet dataset for the ILSVRC challenge. The AlexNet was the first CNN that won the ILSVRC 2012 challenge and was a network with five convolutional

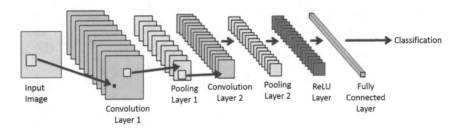

Fig. 2.5 An example of a convolutional neural network (CNN) architecture

Table 2.1 The top 5 error in the classification task using the ImageNet dataset

No.	CNN	Top 5 error (%)
1	AlexNet [35]	16.40
2	ZFNet [88]	11.70
3	VGG Net [68]	7.30
4	GoogLeNet [70]	6.70
5	ResNet [28]	3.57

layers, five max-pooling layers and three fully-connected layers. The ZFNet was then developed that improved the AlexNet by fine-tuning the architecture of the AlexNet. The ILSVRC 2014 challenge winner was the GoogLeNet which used an inception module to remove a large number of parameters from the network for improved efficiency. The current state-of-the-art CNN and the winner of ILSVRC 2015 is the ResNet which introduces the concept of residual net with skip connections and batch normalization with 152 layers in the architecture.

2.2.4 Deep Boltzmann Machine (DBM)

A Deep Boltzmann Machine (DBM) [22, 23, 61] is a type of generative model where the variables with each layer depend on each other conditioned on the neighbouring variables. A DBM is a probablistic graph model containing stacked layers of a Restricted Boltzmann Machine (RBM) where the connections between all the layers are undirected. For a DBM with a visible layer $\mathbf{v}$, three hidden layers $\mathbf{h}_1$, $\mathbf{h}_2$ and $\mathbf{h}_3$, weight matrices $\mathbf{W}_1$, $\mathbf{W}_2$ and $\mathbf{W}_3$, and the partition function Z, the joint probability is given as follows [61]:

$$P(\mathbf{v}) = \sum_{\mathbf{h}_1, \mathbf{h}_2, \mathbf{h}_3} \frac{1}{Z} \exp[\mathbf{v}^T \mathbf{W}_1 \mathbf{h}_1 + \mathbf{h}_1{}^T \mathbf{W}_2 \mathbf{h}_2 + \mathbf{h}_2{}^T \mathbf{W}_3 \mathbf{h}_3]$$

The pre-training for a DBM must be initialized from stacked restricted Boltzmann machines or RBMs, and a discriminative fine tuning is performed using the error back propagation algorithm. A DBM derives a high level representation from the unlabeled data while the labeled data is only used to slightly fine-tune the data. Experimental results on several visual recognition datasets show that the DBM achieves better performance than some other learning methods [22, 61].

2.3 Support Vector Machines

Support Vector Machine (SVM) is one of the popular machine learning methods. The fundamental idea behind SVM is a novel statistical learning theory that was proposed by Vapnik [77]. Unlike traditional methods such as Neural Networks which are based on the empirical risk minimization (ERM), SVM was based on the VC dimension and the structural risk minimization (SRM) [77]. Since its introduction, SVM has been applied to a number of applications ranging from text detection and categorization [32, 67], handwritten character and digit recognition [73], speech verification and recognition [48], face detection and recognition [72], to object detection and recognition [53].

Though SVM achieves better generalization performance compared with many other machine learning technologies, when solving large-scale and complicated problems, the learning process of SVM tends to define a complex decision model due to the increasing number of support vectors. As a result, SVM becomes less efficient due to the expensive computation cost, which involves the inner product computation for a linear SVM and the kernel computation for a kernel SVM for all the support vectors. Many new SVM approaches have been proposed to address the inefficiency problem (i.e. the large number of support vectors) of the conventional SVM [8, 11, 15, 36, 38, 54, 58, 65]. We will discuss these approaches in Sect. 2.3.4 and present a new efficient Support Vector Machine (eSVM) in Sect. 2.3.5 [12–14].

2.3.1 Linear Support Vector Machine

To introduce the linear SVM we will start with outlining the application of SVM to the simplest case of binary classification that is linearly separable. Let the training set be $\{(\mathbf{x}_1, y_1), (\mathbf{x}_2, y_2), \ldots, (\mathbf{x}_l, y_l)\}$, where $\mathbf{x}_i \in \mathbb{R}^n$, $y_i \in \{-1, 1\}$ indicate the two different classes, and l is the number of the training samples. An n dimensional hyperplane that can completely separate the samples may be defined as:

$$\mathbf{w}^t \mathbf{x} + b = 0 \tag{2.1}$$

This hyperplane is defined such that $\mathbf{w}^t \mathbf{x} + b \geq +1$ for the positive samples and $\mathbf{w}^t \mathbf{x} + b \leq -1$ for the negative samples. As mentioned above, SVM searches the optimal separating hyperplane by maximizing the geometric margin. This margin can be maximized as follows:

$$\min_{\mathbf{w}, b} \frac{1}{2} \mathbf{w}^t \mathbf{w}, \\ subject\ to\ \ y_i(\mathbf{w}^t \mathbf{x}_i + b) \geq 1, \quad i = 1, 2, \ldots, l. \tag{2.2}$$

The Lagrangian theory and the Kuhn-Tucker theory are then applied to optimize the functional in Eq. 2.2 with inequality constraints [78]. The optimization process leads to the following quadratic convex programming problem:

$$\max_{\alpha} \sum_{i=1}^{l} \alpha_i - \frac{1}{2} \sum_{i,j=1}^{l} \alpha_i \alpha_j y_i y_j \mathbf{x}_i \mathbf{x}_j$$
$$subject\ to\ \sum_{i=1}^{l} y_i \alpha_i = 0, \tag{2.3}$$
$$\alpha_i \geq 0,\ i = 1, 2, \ldots l$$

From the Lagrangian theory and the Kuhn-Tucker theory, we also have:

$$\mathbf{w} = \sum_{i=1}^{l} y_i \alpha_i \mathbf{x}_i = \sum_{i \in SV} y_i \alpha_i \mathbf{x}_i \tag{2.4}$$

where SV is the set of Support Vectors (SVs), which are the training samples with nonzero coefficients α_i. The decision function of the SVM is therefore derived as follows:

$$f(x) = sign(\mathbf{w}\mathbf{x} + b) = sign(\sum_{i \in SV} y_i \alpha_i \mathbf{x}_i \mathbf{x} + b) \tag{2.5}$$

2.3.2 Soft-Margin Support Vector Machine

In applications with real data, the two classes are usually not completely linearly separable. The soft-margin SVM was then proposed with a tolerance of misclassification error [78]. The fundamental idea of the soft-margin SVM is to maximize the margin of the separating hyperplane while minimizing a quantity proportional to the misclassification errors. To do this, the soft-margin SVM introduces the slack variables $\xi_i \geq 0$ and a regularizing parameter $C > 0$. The soft-margin SVM is defined as follows:

$$\min_{\mathbf{w},b,\xi_i} \frac{1}{2}\mathbf{w}^t\mathbf{w} + C \sum_{i=1}^{l} \xi_i ,$$
$$subject\ to\ y_i(\mathbf{w}^t\mathbf{x}_i + b) \geq 1 - \xi_i , \tag{2.6}$$
$$\xi_i \geq 0,\ i = 1, 2, \ldots, l.$$

Using the Lagrangian theory, its quadratic convex programming program is defined as follows:

$$\max_{\alpha} \sum_{i=1}^{l} \alpha_i - \frac{1}{2} \sum_{i,j=1}^{l} \alpha_i \alpha_j y_i y_j \mathbf{x}_i \mathbf{x}_j$$
$$subject\ to\ \sum_{i=1}^{l} y_i \alpha_i = 0, \tag{2.7}$$
$$0 \leq \alpha_i \leq C,\ i = 1, 2, \ldots l$$

From Eq. 2.6 we can observe that the standard SVM is defined on the trade-off between the least number of misclassified samples ($\min_{\xi_i} C \sum_{i=1}^{l} \xi_i$) and the maximum margin ($\min_{\mathbf{w},b} \frac{1}{2}\mathbf{w}^t\mathbf{w}$) of the separating hyperplane. The decision function of the soft-margin SVM is the same with the linear SVM.

2.3.3 Non-linear Support Vector Machine

The linear SVM and the soft-margin SVM are generally not suitable for complex classification problems which are completely inseparable. Non-linear Support Vector Machine solves the non-linear classification by mapping the data from the input space into a high dimensional feature space using a non-linear transformation Φ : $x_i \rightarrow \phi(x_i)$. Cover's theorem states that if the transformation is nonlinear and the dimensionality of the feature space is high enough, then the input space may be transformed into a new feature space where the data are linearly separable with high probability [77]. This nonlinear transformation is performed in an implicit way through kernel functions [77].

Specifically, the non-linear SVM is defined as follows:

$$
\begin{aligned}
&\min_{\mathbf{w},b,\xi_i} \tfrac{1}{2}\mathbf{w}^t\mathbf{w} + C \sum_{i=1}^{l} \xi_i , \\
&subject\ to\ \ y_i(\mathbf{w}^t\phi(\mathbf{x}_i) + b) \geq 1 - \xi_i , \\
&\xi_i \geq 0, \ \ i = 1, 2, \ldots, l.
\end{aligned}
\tag{2.8}
$$

Its corresponding quadratic convex programming program is as follows:

$$
\begin{aligned}
&\max_{\alpha} \sum_{i=1}^{l} \alpha_i - \tfrac{1}{2} \sum_{i,j=1}^{l} \alpha_i\alpha_j y_i y_j K(\mathbf{x}_i, \mathbf{x}_j) \\
&subject\ to\ \ \sum_{i=1}^{l} y_i\alpha_i = 0, \\
&0 \leq \alpha_i \leq C, \ i = 1, 2, \ldots l
\end{aligned}
\tag{2.9}
$$

where $K(\mathbf{x}_i, \mathbf{x}_j) = \phi(\mathbf{x}_i)^t\phi(\mathbf{x}_j)$ is the kernel function. The decision function of the non-linear SVM is defined as follows:

$$
f(x) = sign(\sum_{i \in SV} y_i\alpha_i K(\mathbf{x}_i, \mathbf{x}) + b)
\tag{2.10}
$$

Typically, there are three types of kernel functions, namely, the polynomial kernel functions, the Gaussian kernel functions, and the Sigmoid functions (though strictly speaking, the Sigmoid functions are not kernel functions).

1. Polynomial: $K(\mathbf{x}_i, \mathbf{x}_j) = (\mathbf{x}_i^t\mathbf{x}_j + 1)^d$
2. Gaussian: $K(\mathbf{x}_i, \mathbf{x}_j) = exp(-\frac{\|\mathbf{x}_i - \mathbf{x}_j\|^2}{2\sigma^2})$
3. Sigmoid: $K(\mathbf{x}_i, \mathbf{x}_j) = tanh(\beta_0\mathbf{x}_i^t\mathbf{x}_j + \beta_1)$

2.3.4 Simplified Support Vector Machines

Although SVM exhibits many theoretical and practical advantages like good generalization performance, the decision function of SVM involves a kernel computation with all Support Vectors (SVs) and thus leads to a slow testing speed. This situation becomes even worse when SVM is applied to large-scale and complicated problems like image search and video retrieval. The decision function becomes over complex due to the large number of SVs and the testing speed is extremely slow because of the expensive kernel computation cost. To address this problem, much research has been carried out to simplify the SVM model and some simplified SVMs have been proposed. In this section, we will first thoroughly analyze the structure and distribution of SVs in the traditional SVM as well as its impact on the computation cost and generalization performance. We will then review some representative simplified SVMs.

Previous research shows that the complexity of a classification model depends on the size of its parameters [1]. A simple model can generate a fast system but has poor accuracy. In contrast, a complex model can reach a higher classification accuracy on the training data but will lead to lower efficiency and poor generalization performance.

From Eq. 2.10, it is observed that the complexity of SVM model depends on the size of SVs, which are defined as a subset of training samples whose corresponding α_i is not equal to zero. According to the Karush-Kuhn-Tucker (KKT) conditions, the optimization problem of the standard SVM defined in Eq. 2.4 should satisfy the following equation:

$$\alpha_i[y_i(\omega^t\phi(x) + b) - 1 + \xi_i] = 0, \ i = 1, 2, \ldots l \qquad (2.11)$$

where $\alpha_i \neq 0$ when $y_i(\omega^t\phi(x) + b) - 1 + \xi_i = 0$. Because of the flexibility of the parameter ξ_i, the probability that $y_i(\omega^t\phi(x) + b) - 1 + \xi_i = 0$ holds is very high, and thus α_i is more likely to get a nonzero value. More specifically, in Eq. 2.11, SVs are those samples between and on the two separating hyperplanes $\omega^t\phi(x) + b = -1$ and $\omega^t\phi(x) + b = 1$ (Fig. 2.6). For a complicated large-scale classification problem, since many misclassified samples exist between these two hyperplanes during training, the size of SVs will be very large and thus an over complex decision model will be generated.

As we mentioned above, the primary impact of an over complex model is on its computation efficiency. From Eq. 2.10, it is observed that the decision function of SVM involves a kernel computation with all SVs. Therefore, the computational complexity of SVM is proportional to the number of SVs. An over complex model contains a large number of SVs and thus its computational cost becomes very expensive. This will lead to a slower classification speed that restricts the application of SVM to real-time applications. Another potential harm of an over complex SVM model is to reduce its generalization performance. SVM is well known for its good generalization performance. Unlike the techniques such as the Artificial Neural

Fig. 2.6 SVM in 2D space
(*Red circles* represent
support vectors)

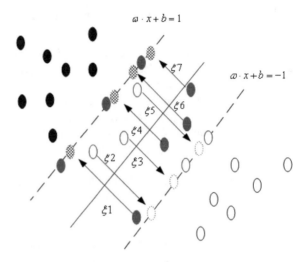

Networks (ANNs) that are based on the Empirical Risk Minimization (ERM) principle, SVM is based on the Structural Risk Minimization (SRM) principle [77]. the SRM principle empowers SVM with good generalization performance by keeping a balance between seeking the best classifier using the training data and avoiding overfitting during the learning process. However, an over complex model is likely to break this balance and increases the risk of overfitting and thus reduces its generalization performance.

Burges [8] proposed a method that computes an approximation to the decision function in terms of a reduced set of vectors to reduce the computational complexity of the decision function by a factor of ten. The method was then applied to handwritten digits recognition [65] and face detection [58]. However, this method not only reduces the classification accuracy but also slows down the training speed due to the higher computational cost for the optimal approximation. A Reduced Support Vector Machine (RSVM) was then proposed as an alternative to the standard SVM [36, 38]. A nonlinear kernel was generated based on a separating surface (decision function) by solving a smaller optimization problem using a subset of the training samples. The RSVM successfully reduced the model's complexity but it also decreased the classification rate. Furthermore, a new SVM, named υ-SVM, was proposed [15]. The relationship among the parameter υ, the number of support vectors, and the classification error was thoroughly discussed. However, this method would reduce the generalization performance when the parameter υ is too small. Other simplified support vector machine models are introduced in [11, 54]. To summarize, most of these simplified SVMs are able to reduce the computational cost but often at the expense of accuracy.

2.3.5 Efficient Support Vector Machine

We now introduce an efficient Support Vector Machine (eSVM), which significantly improves the computational efficiency of the traditional SVM without sacrificing its generalization performance [13]. The eSVM has been successfully applied to eye detection [14]. Motivated by the above analysis that it is the flexibility of the parameter ξ_i that leads to the large number of support vectors, the eSVM implements the second optimization of Eq. 2.8 as follows:

$$\min_{\omega,b,\xi} \tfrac{1}{2}\omega^t\omega + C\xi \,,$$
$$subject\ to\ \ y_i(\omega^t\phi(x_i)+b) \geq 1\,,\ i \in V - MV$$
$$y_i(\omega^t\phi(x_i)+b) \geq 1-\xi\,,\ i \in MV\,,\ \xi \geq 0 \tag{2.12}$$

where M is the set of the misclassified samples in the traditional SVM and V is the set of all training samples. Its dual quadratic convex programming problem is:

$$\max_{\alpha} \sum_{i \in V} \alpha_i - \tfrac{1}{2} \sum_{i,j \in V} \alpha_i \alpha_j y_i y_j K(x_i, x_j)$$
$$subject\ to\ \ \sum_{i \in V} y_i \alpha_i = 0,\ \left(\sum_{i \in MV} \alpha_i\right) \leq C, \tag{2.13}$$
$$\alpha_i \geq 0,\ i \in V$$

Note that instead of the flexibility of the slack variables in Eq. 2.8, we set these slack variables to a fixed value in Eq. 2.12. Now the new KKT conditions of Eq. 2.12 become:

$$\alpha_i[y_i(\omega^t\phi(x)+b) - 1] = 0,\ i \in V - MV$$
$$\alpha_i[y_i(\omega^t\phi(x)+b) - 1 + \xi] = 0,\ i \in MV \tag{2.14}$$

According to Eq. 2.14, $\alpha_i \neq 0$ when $y_i(\omega^t\phi(x)+b) - 1 = 0$, $i \in V - MV$ or $y_i(\omega^t\phi(x)+b) - 1 + \xi = 0$, $i \in MV$. The support vectors are those samples on the two separating hyperplanes $\omega^t\phi(x) + b = -1$ and $\omega^t\phi(x) + b = 1$ and the mis-classified samples farthest away from the hyperplanes (Fig. 2.7). As a result, the number of support vectors is much less than those defined by Eq. 2.11.

Compared with the traditional SVM, which is defined on the trade-off between the least number of the misclassified samples ($\min_{\xi} C \sum_{i=1}^{l} \xi_i$) and the maximum margin ($\min_{\omega,b} \tfrac{1}{2}\omega^t\omega$) of the two separating hyperplanes, the eSVM is defined on the trade-off between the minimum margin of the misclassified samples ($\min_{\xi} C\xi$) and the maximum margin ($\min_{\omega,b} \tfrac{1}{2}\omega^t\omega$) between the two separating hyperplanes. For complicated classification problems, the traditional SVM builds up a complex SVM model in pursuit of the least number of misclassified samples to some extend. According to SRT, it will increase the risk of overfitting on the training samples and thus

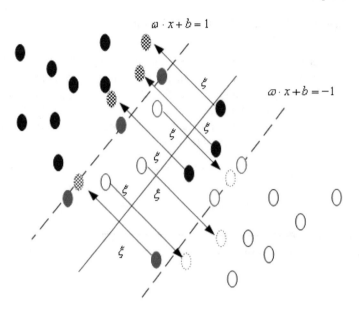

$$\omega \cdot x + b = 1$$

$$\omega \cdot x + b = -1$$

Fig. 2.7 eSVM in 2D space (*Red circles* represent support vectors)

reduces its generalization performance. The eSVM pursues the minimal margin of misclassified samples and its decision function is more concise. Therefore, the eSVM can be expected to achieve higher classification accuracy than the traditional SVM.

2.3.6 Applications of SVM

This section presents a survey of the applications of SVM to image search and video retrieval. Rapidly increasing use of smart phones and significantly reduced storage cost have resulted in the explosive growth of digital images and videos over the internet. As a result, image or video based search engines are in an urgent demand to find similar images or videos from a huge image or video database. Over the last decade many learning based image search and video retrieval techniques have been presented. Among them SVM as a powerful learning tool is widely used.

Chang and Tong [75] presented an image search method using a so called support vector machine active learning (SVM_{Active}). SVM_{Active} combines active learning with support vector machine. Intuitively, SVM_{Active} works by combining three ideas: (1) SVM_{Active} regards the task of learning a target concept as one of learning an SVM binary classifier. An SVM captures the query concept by separating the relevant images from irrelevant ones with a hyperplane in a projected space, usually a very high-dimensional one. The projected points on one side of the hyperplane are considered relevant to the query concept and the rest irrelevant; (2) SVM_{Active} learns the classifier quickly via active learning. The active part of SVM_{Active} selects the most informative instances with which to train the SVM classifier. This step ensures fast convergence to the query concept in a small number of feedback rounds; (3) once the classifier is trained, SVM_{Active} returns the top-k most relevant images. These are the k images farthest from the hyperplane on the query concept side. SVM_{Active} needs at least one positive and one negative example to start its learning process. Two seeding methods were also presented: one by MEGA and one by keywords. To make both concept-learning and image retrieval efficient, a multi-resolution image-feature extractor and a high-dimensional indexer were also applied. Experiments ran on three real-world image data sets that were collected from Corel Image CDs and the internet (http://www.yestart.com/pic/). Those three data sets contain a four-category, a ten-category, and a fifteen-category image sets, respectively. Each category consists of 100–150 images. Experimental results show that SVM_{Active} achieves significantly higher search accuracy than the traditional query refinement schemes.

Guo et al. [25] presented a novel metric to rank the similarity for texture image search. This metric was named distance from boundary (DFB), in which the boundary is obtained by SVM. In conventional texture image retrieval, the Euclidean or the Mahalanobis distances between the images in the database and the query image are calculated and used for ranking. The smaller the distance, the more similar the pattern to the query. But this kind of metric has some limitations: (1) the retrieval results corresponding to different queries may be much different although they are visually similar; (2) the retrieval performance is sensitive to the sample topology; (3) the retrieval accuracy is low. The basic idea of the DFB is to learn a non-linear boundary that separates the similar images with the query image from the remaining ones. SVM is applied to learn this non-linear boundary due to its generalization performance. Compared with the traditional similarity measure based ranking, the DFB method has three advantages: (1) the retrieval performance is relatively insensitive to the sample distribution; (2) the same results can be obtained with respect to different (but visually similar) queries; (3) the retrieval accuracy is improved compared with traditional methods. Experiments on the Brodatz texture image database [7] with 112 texture classes show the effectiveness of the DFB method.

Recently the interactive learning mechanism was introduced to image search. The interactive learning involves a relevance feedback (RF) that is given by users to indicate which images they think are relevant to the query. Zhang et al. [89] presented an SVM based relevance feedback for image search. Specifically, during the process of relevance feedback, users can mark an image as either relevant or irrelevant. Given the top N_{RF} images in the result as the training data, a binary classifier can be learned using an SVM to properly represent the user's query. An SVM is chosen here due to its generalization performance. Using this binary classifier, other images can be classified into either the relevance class or the irrelevance class in terms of the distance from each image to the separating hyperplane. Obviously, in the first learning iteration, both marked relevant samples and unmarked irrelevant samples are all close to the query. Such samples are very suitable to construct the SVM classifier because support vectors are just those that lie on the separating margin while other samples far away from the hyperplane will contribute nothing to the classifier. In the following iterations, more relevant samples fed back by users can be used to refine the classifier. Experiments were performed on a database that consists of 9,918 images from the Corel Photo CD (http://www.yestart.com/pic/). Five iterations were carried out to refine the SVM classifier, with each iteration allowing users to mark top 100 ($N_{RF} = 100$) images as feedback. A Gaussian kernel was chosen for the SVM. Experimental results show that both the recall rate and the precision rate are improved as the learning iteration progresses, and finally it reaches a satisfactory performance.

Hong et al. [31] presented a method that utilized SVM to update the preference weights (through the RF) that are used to evaluate the similarity between the relevant images. The similarity between two images – the query image and the searched image—is defined by summing the distances of individual features with fixed preference weights. The weights can be updated through the RF to reach better search performance. In [31], SVM was applied to perform non-linear classification on the positive and negative feedbacks through the RF. The SVM learning results are then utilized to automatically update the preference weights. Specifically, once the SVM separating hyperplane was trained, the distance between a feedback sample and the separating hyperplane indicates that how much this sample belonging to the assigned class is differentiated from the non-assigned one. In other words, the farther the positive sample feedbacks from the hyperplane, the more distinguishable they are from the negative samples. Therefore, those samples should be assigned a larger weight compared with other samples. In [31], the preference weight is set linearly proportional to the distance between the sample and the separating hyperplane. Experiments were performed on the COREL dataset (http://www.yestart.com/pic/), which contains 17,000 images. A polynomial kernel function with $d = 1$ was used for the SVM learning. The preference weights were normalized to the range [10, 100]. Experimental results show improved accuracy over other RF based methods.

Li et al. [37] presented a multitraining support vector machine (MTSVM) to further improve the SVM based RF for image retrieval. The MTSVM is based on the observation that (1) the success of the co-training model augments the labeled examples with unlabeled examples in information retrieval; (2) the advances in the random subspace method overcomes the small sample size problem. With the incorporation of the SVM and the multitraining model, the unlabeled examples can generate new informative training examples for which the predicted labels become more accurate. Therefore, the MTSVM method can work well in practical situations. In the MTSVM learning model, the majority voting rule (MVR) [34] was chosen as the similarity measure in combining individual classifiers since every single classifier has its own distinctive ability to classify relevant and irrelevant samples. Experiments were carried out upon a subset of images from the Corel Photo Gallery (http://www.yestart.com/pic/). This subset consists of about 20,000 images of very diverse subject matters for which each image was manually labeled with one of the 90 concepts. Initially, 500 queries were randomly selected, and the program autonomously performs a RF with the top five most relevant images (i.e., images with the same concept as the query) marked as positive feedback samples within the top 40 images. The five negative feedback samples are marked in a similar fashion. The procedure is chosen to replicate a common working situation where a user would not label many images for each feedback iteration. Experimental results shown that MTSVM consistently improved the performance over conventional SVM-based RFs in terms of precision and standard deviation.

With the observation of the success application of RF and SVM to image retrieval, Yazdi et al. [87] applied RF and SVM to video retrieval. The proposed method consists of two major steps: key frame extraction and video shot retrieval. A new frame extraction method was presented using a hierarchical approach based on clustering. Using this method, the most representative key frame was then selected for each video shot. The video retrieval was based on an SVM based RF that was capable of combining both low-level features and high-level concepts. The low-level features are the visual image features such as color and texture, while the high-level concepts are the user's feedback through RF. The video database was finally classified into groups of relevant and irrelevant using this SVM classifier. The proposed method was validated on a video database with 800 shots from Trecvid2001 (http:// www.open-video.org) and home videos. The video shots database includes airplanes, jungles, rivers, mountains, wild life, basketball, roads, etc. A total of 100 random queries were selected and judgements on the relevance of each shot to each query shot were evaluated. Different kernels for SVM-based learning in RF module were used. Experimental results show that SVM with the Gaussian function as kernel has better performance than the linear or polynomial kernel. The final experimental results show the improved performance after only a few RF iterations.

More applications of SVM to image search and video retrieval can be found in [10, 30, 52, 62, 71, 86].

2.4 Other Popular Kernel Methods and Similarity Measures

Kernel methods stress the nonlinear mapping from an input space to a high-dimensional feature space [18, 29, 63, 66]. The theoretical foundation for implementing such a nonlinear mapping is the Cover's theorem on the separability of patterns: "A complex pattern-classification problem cast in a high-dimensional space nonlinearly is more likely to be linearly separable than in a low-dimensional space" [26]. Support Vector Machine (SVM) [18, 78], which defines an optimal hyperplane with maximal margin between the patterns of two classes in the feature space mapped nonlinearly from the input space, is a kernel method. Kernel methods have been shown more effective than the linear methods for image classification [3, 16, 50, 64]. Being linear in the feature space, but nonlinear in the input space, kernel methods thus are capable of encoding the nonlinear interactions among patterns. Representative kernel methods, such as kernel PCA [64] and kernel FLD [3, 16, 50], overcome many limitations of the corresponding linear methods by nonlinearly mapping the input space to a high-dimensional feature space. Scholkopf et al. [64] showed that kernel PCA outperforms PCA using an adequate non-linear representation of input data. Yang et al. [85] compared face recognition performance using kernel PCA and the Eigenfaces method. The empirical results showed that the kernel PCA method with a cubic polynomial kernel achieved the lowest error rate. Moghaddam [51] demonstrated that kernel PCA with Gaussian kernels performed better than PCA for face recognition. Some representative kernel methods include kernel discriminant analysis [56, 84], kernel-based LDA [47], localized kernel eigenspaces [24], sparse kernel feature extraction [19], and multiple kernel learning algorithm [79, 80, 82].

Further research shows that new kernel methods with unconventional kernel models are able to improve pattern recognition performance [40]. One such kernel method is the multi-class Kernel Fisher Analysis (KFA) method [40]. The KFA method extends the two-class kernel Fisher methods [16, 50] by addressing multi-class pattern recognition problems, and it improves upon the traditional Generalized Discriminant Analysis (GDA) method [3] by deriving a unique solution. As no theoretical guideline is available in choosing a right kernel function for a particular application and the flexibility of kernel functions is restricted by the Mercer's conditions, one should investigate new kernel functions and new kernel models for improving the discriminatory power of kernel methods. The fractional power polynomial models have been shown to be able to improve image classification performance when integrated with new kernel methods [39, 40]. A fractional power polynomial, however, does not necessarily define a kernel function, as it might not define a positive semidefinite Gram matrix. Hence, a fractional power polynomial is called a model rather than a kernel function.

Similarity measures play an essential role in determining the performance of different learning and recognition methods [9, 43, 44, 46, 74]. Some image classification methods, such as the Eigenfaces method [33, 76], often apply the whitened cosine similarity measure for achieving good classification performance [6, 55]. Other methods, such as the Fisherfaces method [4, 20, 69], however, often utilize the cosine similarity measure for improving image classification performance [45, 46]. Further research reveals why the whitened cosine similarity measure achieves good image classification performance for the Principal Component Analysis (PCA) based methods [41, 42]. In addition, new similarity measures, such as the PRM Whitened Cosine (PWC) similarity measure and the Within-class Whitened Cosine (WWC) similarity measure, for further improving image classification performance have been presented [41]. The reason why the cosine similarity measure boosts the image classification performance for the discriminant analysis based methods has been discovered due to its theoretical roots in the Bayes decision rule for minimum error [44]. Furthermore, some inherent challenges of the cosine similarity, such as its inadequacy in addressing distance and angular measures, have been investigated. And finally a new similarity measure, which overcomes the inherent challenges by integrating the absolute value of the angular measure and the l_p norm of the distance measure, is presented for further enhancing image classification performance [44].

2.5 Conclusion

We have reviewed in this chapter some representative learning and recognition methods that have broad applications in intelligent image search and video retrieval. In particular, we first discuss some popular deep learning methods, such as the feedforward deep neural networks [35, 49], the deep autoencoders [2, 81], the convolutional neural networks [23], and the Deep Boltzmann Machine (DBM) [22, 61]. We then discuss one of the popular machine learning methods, namely, Support Vector Machine (SVM) [77]. Specifically we review the linear support vector machine [78], the soft-margin support vector machine [78], the non-linear support vector machine [77], the simplified support vector machine [11, 54], the efficient Support Vector Machine (eSVM) [13, 14], and the applications of SVM to image search and video retrieval [25, 37, 52, 62, 75, 87]. We finally briefly review some other popular kernel methods and new similarity measures [40, 41, 44].

References

1. Alpaydin, E.: Introduction to machine learning. The MIT Press, Cambridge (2010)
2. Baldi, P.: Autoencoders, unsupervised learning, and deep architectures. J. Mach. Learn. Res. (Proceedings of ICML unsupervised and transfer learning) **27**, 37–50 (2011)
3. Baudat, G., Anouar, F.: Generalized discriminant analysis using a kernel approach. Neural Comput. **12**(10), 2385–2404 (2000)
4. Belhumeur, P.N., Hespanha, J.P., Kriegman, D.J.: Eigenfaces vs. Fisherfaces: recognition using class specific linear projection. IEEE Trans. Pattern Anal. Mach. Intell. **19**(7), 711–720 (1997)
5. Bengio, Y., Yao, L., Alain, G., Vincent, P.: Generalized denoising auto-encoders as generative models. Adv, Neural Inf. Process. Syst. **26**, 899–907 (2013)
6. Beveridge, J., Givens, G., Phillips, P., Draper, B.: Factors that influence algorithm performance in the face recognition grand challenge. Comput. Vis. Image Underst. **113**(6), 750–762 (2009)
7. Brodatz, P.: Textures: a photographic album for artists and designers. Dover, New York (1966)
8. Burges, C.: Simplified support vector decision rule. In: Proceedings of the Thirteenth International Conference on Machine Learning (ICML'96), Bari, Italy, July 3–6, 1996 (1996)
9. Chambon, S., Crouzil, A.: Similarity measures for image matching despite occlusions in stereo vision. Pattern Recognit. **44**(9), 2063–2075 (2011)
10. Chapelle, O., Haffner, P., Vapnik, V.N.: Support vector machines for histogram-based image classification. IEEE Trans. Neural Netw. **10**(5), 1055–1064 (1999)
11. Chen, J., Chen, C.: Reducing SVM classification time using multiple mirror classifers. IEEE Trans. Syst. Man Cybern. **34**(2), 1173–1183 (2004)
12. Chen, S., Liu, C.: Eye detection using color information and a new efficient SVM. In: IEEE Fourth International Conference on Biometrics: Theory, Applications, and Systems (BATS'10), Washington DC, USA (2010)

13. Chen, S., Liu, C.: A new efficient SVM and its application to a real-time accurate eye localization system. In: International Joint Conference on Neural Networks, San Jose, California, USA (2011)
14. Chen, S., Liu, C.: Eye detection using discriminatory haar features and a new efficient SVM. Image Vis. Comput. **33**(c), 68–77 (2015)
15. Chen, P., Lin, C., Scholkopf, B.: A tutorial on υ-support vector machines. Appl. Stoch. Models Bus. Ind. **21**, 111–136 (2005)
16. Cooke, T.: Two variations on Fisher's linear discriminant for pattern recognition. IEEE Trans. Pattern Anal. Mach. Intell. **24**(2), 268–273 (2002)
17. Corso, J.J., Alahi, A., Grauman, K., Hager, G.D., Morency, L.P., Sawhney, H., Sheikh, Y.: Video Analysis for Bodyworn Cameras in Law Enforcement. The Computing Community Consortium whitepaper (2015)
18. Cristianini, N., Shawe-Taylor, J.: An Introduction to Support Vector Machines and Other Kernel-Based Learning Methods. Cambridge University Press, Cambridge (2000)
19. Dhanjal, C., Gunn, S., Shawe-Taylor, J.: Efficient sparse kernel feature extraction based on partial least squares. IEEE Trans. Pattern Anal. Mach. Intell. **31**(8), 1347–1361 (2009)
20. Etemad, K., Chellappa, R.: Discriminant analysis for recognition of human face images. J. Opt. Soc. Am. A **14**, 1724–1733 (1997)
21. Glorot, X., Bordes, A., Bengio, Y.: Deep sparse rectifier neural networks. In: Aistats, vol. 15, p. 275 (2011)
22. Goodfellow, I., Mirza, M., Courville, A., Bengio, Y.: Multi-prediction deep boltzmann machines. In: Advances in Neural Information Processing Systems, pp. 548–556 (2013)
23. Goodfellow, I., Bengio, Y., Courville, A.: Deep learning (2016). URL http://www.deeplearningbook.org (Book in preparation for MIT Press)
24. Gundimada, S., Asari, V.: Facial recognition using multisensor images based on localized kernel eigen spaces. IEEE Trans. Image Process. **18**(6), 1314–1325 (2009)
25. Guo, G., Zhang, H.J., Li, S.Z.: Distance-from-boundary as a metric for texture image retrieval. In: IEEE International Conference on Acoustics. Speech, and Signal Processing, vol. 3, pp. 1629–1632, Washington DC, USA (2001)
26. Haykin, S.: Neural Networks — A Comprehensive Foundation. Macmillan College Publishing Company, Inc., New York (1994)
27. He, K., Zhang, X., Ren, S., Sun, J.: Delving deep into rectifiers: Surpassing human-level performance on imagenet classification. In: 2015 IEEE International Conference on Computer Vision (ICCV), pp. 1026–1034 (2015)
28. He, K., Zhang, X., Ren, S., Sun, J.: Deep residual learning for image recognition. In: The IEEE Conference on Computer Vision and Pattern Recognition (CVPR) (2016)
29. Herbrich, R.: Learning Kernel Classifiers: Theory and Algorithms. MIT Press, Cambridge (2002)
30. Hoi, C.H., Chan, C.H., Huang, K., Lyu, M.R., King, I.: Biased support vector machine for relevance feedback in image retrieval. In: 2004 IEEE International Joint Conference on Neural Networks, vol. 4, pp. 3189–3194 (2004)
31. Hong, P., Tian, Q., Huang, T.S.: Incorporate support vector machines to content-based image retrieval with relevance feedback. In: 2000 International Conference on Image Processing, vol. 3, pp. 750–753 (2000)
32. Joachims, T.: Text categorization with support vector machines: learning with many relevant features. In: 10th European Conference on Machine Learning (1999)
33. Kirby, M., Sirovich, L.: Application of the Karhunen-Loeve procedure for the characterization of human faces. IEEE Trans. Pattern Anal. Mach. Intell. **12**(1), 103–108 (1990)
34. Kittler, J., Hatef, M., Duin, R., Matas, J.: On combining classifiers. IEEE Trans. Pattern Anal. Mach. Intell. **20**(3), 226–239 (1998)
35. Krizhevsky, A., Sutskever, I., Hinton, G.E.: Imagenet classification with deep convolutional neural networks. In: Pereira, F., Burges, C.J.C., Bottou, L., Weinberger, K.Q. (eds.) Advances in Neural Information Processing Systems 25, pp. 1097–1105 (2012)

36. Lee, Y., Mangasarian, O.: RSVM: Reduced support vector machines. In: The First SIAM International Conference on Data Mining (2001)
37. Li, J., Allinson, N., Tao, D., Li, X.: Multitraining support vector machine for image retrieval. IEEE Trans. Image Process. **15**(11), 3597–3601 (2006)
38. Lin, K., Lin, C.: A study on reduced support vector machine. IEEE Trans. Neural Netw. **14**(6), 1449–1559 (2003)
39. Liu, C.: Gabor-based kernel PCA with fractional power polynomial models for face recognition. IEEE Trans. Pattern Anal. Mach. Intell. **26**(5), 572–581 (2004)
40. Liu, C.: Capitalize on dimensionality increasing techniques for improving face recognition grand challenge performance. IEEE Trans. Pattern Anal. Mach. Intell. **28**(5), 725–737 (2006)
41. Liu, C.: The Bayes decision rule induced similarity measures. IEEE Trans. Pattern Anal. Mach. Intell. **29**(6), 1086–1090 (2007)
42. Liu, C.: Clarification of assumptions in the relationship between the bayes decision rule and the whitened cosine similarity measure. IEEE Trans. Pattern Anal. Mach. Intell. **30**(6), 1116–1117 (2008)
43. Liu, C.: Effective use of color information for large scale face verification. Neurocomputing **101**, 43–51 (2013)
44. Liu, C.: Discriminant analysis and similarity measure. Pattern Recognit. **47**(1), 359–367 (2014)
45. Liu, Z., Liu, C.: Fusion of color, local spatial and global frequency information for face recognition. Pattern Recognit. **43**(8), 2882–2890 (2010)
46. Liu, C., Mago, V. (eds.): Cross Disciplinary Biometric Systems. Springer, New York (2012)
47. Liu, X., Chen, W., Yuen, P., Feng, G.: Learning kernel in kernel-based LDA for face recognition under illumination variations. IEEE Signal Process. Lett. **16**(12), 1019–1022 (2009)
48. Ma, C., Randolph, M., Drish, J.: A support vector machines-based rejection technique for speech recognition. In: IEEE International Conference on Acoustics, Speech, and Signal Processing, vol. 1, pp. 381–384 (2001)
49. Maas, A.L., Hannun, A.Y., Ng, A.Y.: Rectifier nonlinearities improve neural network acoustic models. In: Proceedings of ICML, vol. 30 (2013)
50. Mika, S., Ratsch, G., Weston, J., Scholkopf, B., Mller, K.R.: Fisher discriminant analysis with kernels. In: Hu, Y.H., Larsen, J., Wilson, E., Douglas, S. (eds.) Neural Networks for Signal Processing IX, pp. 41–48. IEEE (1999)
51. Moghaddam, B.: Principal manifolds and probabilistic subspaces for visual recognition. IEEE Trans. Pattern Anal. Mach. Intell. **24**(6), 780–788 (2002)
52. Nagaraja, G., Murthy, S.R., Deepak, T.: Content based video retrieval using support vector machine classification. In: 2015 IEEE International Conference on Applied and Theoretical Computing and Communication Technology, pp. 821–827 (2015)
53. Nakajima, C., Pontil, M., Poggio, T.: People recognition and pose estimation in image sequences. In: IEEE International Joint Conference on Neural Networks, vol. 4, pp. 189–194 (2000)
54. Nguyen, D., Ho, T.: An efficient method for simplifying support vector machines. In: International Conference on Machine Learning, Bonn, Germany (2005)
55. OToole, A.J., Phillips, P.J., Jiang, F., Ayyad, J., Penard, N., Abdi, H.: Face recognition algorithms surpass humans matching faces across changes in illumination. IEEE Trans. Pattern Anal. Mach. Intell. **29**(9), 1642–1646 (2007)
56. Pekalska, E., Haasdonk, B.: Kernel discriminant analysis for positive definite and indefinite kernels. IEEE Trans. Pattern Anal. Mach. Intell. **31**(6), 1017–1032 (2009)
57. Ranzato, M.A., Szummer, M.: Semi-supervised learning of compact document representations with deep networks. In: Proceedings of the 25th International Conference on Machine Learning, ICML '08, pp. 792–799 (2008)
58. Romdhani, S., Torr, B., Scholkopf, B., Blake, A.: Computationally efficient face detection. In: IEEE International Conference on Computer Vision (2001)
59. Rumelhart, D.E., Hinton, G.E., Williams, R.J.: Learning representations by back-propagating errors. Nature **323**(6088), 533–536 (1986)

60. Russakovsky, O., Deng, J., Su, H., Krause, J., Satheesh, S., Ma, S., Huang, Z., Karpathy, A., Khosla, A., Bernstein, M., Berg, A.C., Fei-Fei, L.: ImageNet large scale visual recognition challenge. Int. J. Comput. Vis. (IJCV) **115**(3), 211–252 (2015). doi:10.1007/s11263-015-0816-y
61. Salakhutdinov, R., Hinton, G.: An efficient learning procedure for deep boltzmann machines. Neural Comput. **24**(8), 1967–2006 (2012)
62. Santhiya, G., Singaravelan, S.: Multi-SVM for enhancing image search. Int. J. Sci. Eng. Res. **4**(6) (2013)
63. Scholkopf, B., Smola, A.: Learning with Kernels: Support Vector Machines, Regularization, Optimization and Beyond. MIT Press, Cambridege (2002)
64. Scholkopf, B., Smola, A., Muller, K.: Nonlinear component analysis as a kernel eigenvalue problem. Neural Comput. **10**, 1299–1319 (1998)
65. Scholkopf, B., Mika, S., Burges, C., Knirsch, P., Muller, K., Ratsch, G., Smola, A.: Input space versus feature space in kernel-based methods. IEEE Trans. Neural Netw. **10**(5), 1000–1017 (1999)
66. Shawe-Taylor, J., Cristianini, N.: Kernel Methods for Pattern Analysis. Cambridge University Press, Cambridge (2004)
67. Shin, C., Kim, K., Park, M., Kim, H.: Support vector machine-based text detection in digital video. In: IEEE Workshop on Neural Networks for Signal Processing (2000)
68. Simonyan, K., Zisserman, A.: Very deep convolutional networks for large-scale image recognition. arXiv preprint arXiv:1409.1556 (2014)
69. Swets, D.L., Weng, J.: Using discriminant eigenfeatures for image retrieval. IEEE Trans. Pattern Anal. Mach. Intell. **18**(8), 831–836 (1996)
70. Szegedy, C., Liu, W., Jia, Y., Sermanet, P., Reed, S., Anguelov, D., Erhan, D., Vanhoucke, V., Rabinovich, A.: Going deeper with convolutions. In: Proceedings of the IEEE Conference on Computer Vision and Pattern Recognition, pp. 1–9 (2015)
71. Tao, D., Tang, X., Li, X., Wu, X.: Asymmetric bagging and random subspace for support vector machines-based relevance feedback in image retrieval. IEEE Trans. Pattern Anal. Mach. intell. **28**(7), 1088–1099 (2006)
72. Tefas, A., Kotropoulos, C., Pitas, I.: Using support vector machines to enhance the performance of elastic graph matching for frontal face authentication. IEEE Trans. Pattern Anal. Mach. Intell. **23**(7), 735–746 (2001)
73. Teow, L., Loe, K.: Robust vision-based features and classification schemes for off-line handwritten digit recognition. Pattern Recognit. (2002)
74. Thung, K., Paramesran, R., Lim, C.: Content-based image quality metric using similarity measure of moment vectors. Pattern Recognit. **45**(6), 2193–2204 (2012)
75. Tong, S., Chang, E.: Support vector machine active learning for image retrieval. In: the Ninth ACM International Conference on Multimedia, pp. 107–118 (2001)
76. Turk, M., Pentland, A.: Eigenfaces for recognition. J. Cogn. Neurosci. **13**(1), 71–86 (1991)
77. Vapnik, V.: The Nature of Statistical Learning Theory. Springer, New York, NY (1995)
78. Vapnik, Y.N.: The Nature of Statistical Learning Theory, second edn. Springer, New York (2000)
79. Varma, M., Babu, B.: More generality in efficient multiple kernel learning. In: Proceedings of the International Conference on Machine Learning, Montreal, Canada (2009)
80. Vedaldi, A., Gulshan, V., Varma, M., Zisserman, A.: Multiple kernels for object detection. In: Proceedings of the International Conference on Computer Vision, Kyoto, Japan (2009)
81. Vincent, P., Larochelle, H., Lajoie, I., Bengio, Y., Manzagol, P.A.: Stacked denoising autoencoders: learning useful representations in a deep network with a local denoising criterion. J. Mach. Learn. Res. **11**, 3371–3408 (2010)
82. Wang, Z., Chen, S., Sun, T.: Multik-MHKS: a novel multiple kernel learning algorithm. IEEE Trans. Pattern Anal. Mach. Intell. **30**(2), 348–353 (2008)
83. Wright, H.: Video Analysis for Body-worn Cameras in Law Enforcement (2015). http://www.cccblog.org/2015/08/06/video-analysis-for-body-worn-cameras-in-law-enforcement

84. Xie, C., Kumar, V.: Comparison of kernel class-dependence feature analysis (KCFA) with kernel discriminant analysis (KDA) for face recognition. In: Proceedings of IEEE on Biometrics: Theory, Application and Systems (2007)
85. Yang, M.H., Ahuja, N., Kriegman, D.: Face recognition using kernel Eigenfaces. In: Proc. IEEE International Conference on Image Processing, Vancouver, Canada (2000)
86. Yang, J., Yan, R., Hauptmann, A.G.: Cross-domain video concept detection using adaptive svms. In: 15th ACM International Conference on Multimedia, pp. 188–197 (2007)
87. Yazdi, H.S., Javidi, M., Pourreza, H.R.: SVM-based relevance feedback for semantic video retrieval. Int. J. Signal Imaging Syst. Eng. **2**(3), 99–108 (2009)
88. Zeiler, M.D., Fergus, R.: Visualizing and understanding convolutional networks. In: European Conference on Computer Vision, pp. 818–833. Springer, New York (2014)
89. Zhang, L., Lin, F., Zhang, B.: Support vector machine learning for image retrieval. In: 2001 International Conference on Image Processing, vol. 2, pp. 721–724 (2001)

Chapter 3
Improved Soft Assignment Coding for Image Classification

Qingfeng Liu and Chengjun Liu

Abstract Feature coding plays an important role in image classification, and the soft-assignment coding (SAC) method is popular in many practical applications due to its conceptual simplicity and computational efficiency. The SAC method, however, fails to achieve the optimal image classification performance when compared with the recently developed coding methods, such as the sparse coding and the locality-constrained linear coding methods. This chapter first analyzes the SAC method from the perspective of kernel density estimation, and then presents an improved soft-assignment coding (ISAC) method that enhances the image classification performance of the SAC method and keeps its simplicity and efficiency. Specifically, the ISAC method introduces two enhancements, namely, the thresholding normalized visual word plausibility (TNVWP) and the power transformation method. These improvements are further shown to establish the connection between the proposed ISAC method and the Vector of Locally Aggregated Descriptors (VLAD) coding method. Experiments on four representative datasets (the UIUC sports event dataset, the scene 15 dataset, the Caltech 101 dataset, and the Caltech 256 dataset) show that the proposed ISAC method achieves competitive results to and even better results than some popular image classification methods without sacrificing much computational efficiency.

3.1 Introduction

Image classification is a challenging issue in computer vision and pattern recognition due to the complexity and the variability of the image contents. Among the many approaches proposed, the image classification system based on the bag-of-visual-words model generally consists of five steps. First, the local features are extracted from the image and represented as vectors (local feature descriptors). Second, the

Q. Liu (✉) · C. Liu
New Jersey Institute of Technology, Newark, NJ 07102, USA
e-mail: ql69@njit.edu

C. Liu
e-mail: cliu@njit.edu

© Springer International Publishing AG 2017
C. Liu (ed.), *Recent Advances in Intelligent Image Search and Video Retrieval*,
Intelligent Systems Reference Library 121, DOI 10.1007/978-3-319-52081-0_3

codebook that is composed of visual words is learned from the local feature descriptors. Third, each local feature is encoded using some feature coding methods. And then all the local feature codings in the image are pooled to form a vector that is used to represent the image. Finally, the image vector is input to a image classification system (e.g., support vector machine, random forest, etc.) for classification. Among all these steps, the feature coding and pooling steps are often tied closely and form the core components with respect to the classification performance [1–3]. The bag-of-features (BoF) method [4] and its variants, such as the soft-assignment coding (sometimes called the codeword uncertainty, or the soft voting) [1, 5–7], the sparse coding [8], and the locality-constrained linear coding [9], are among the most successful image classification methods that achieve the state-of-the-art performance on many benchmark datasets [10, 11].

The BoF framework typically extracts local features using the SIFT descriptor and groups them by the k-means methods. Then each local feature is mapped into the nearest cluster and results in an indicator vector with only one non-zero elements. Such a hard coding paradigm, however, produces a large quantization error and fails to model the visual word ambiguity [5]. Thus, the soft-assignment coding is proposed to alleviate the drawback of the hard coding method and assign the membership weights to all the clusters where the local feature belongs. However, despite its efficiency and appealing theory, the soft-assignment coding method is inferior to its counterparts, such as the sparse coding approach [8], and the locality-constrained linear coding method [9], in terms of classification performance.

Recently, the Fisher vector method, the Vector of Locally Aggregated Descriptors (VLAD) method [12–14], and their variants [15] demonstrate some promising results in image classification [1], including some large scale experiments [14]. In essence, the power of these methods comes from the combination of the generative model and the discriminative model [16–18]. For example, the Fisher vector method uses the Gaussian mixture model (GMM) to represent the data (the generative model), and then the Fisher kernel based support vector machine for classification (the discriminative model). The key process is that the Fisher kernel is decomposed into the inner product of the mapping functions explicitly to obtain the new image representation.

The success of the Fisher vector method and its variants motivates us to develop an improved soft-assignment coding by means of kernel density estimation in order to harness the power of probability modeling. In this chapter, we argue that the traditional soft-assignment coding, which estimates the distribution over the visual words given the local features, is often not a reliable estimation and not suitable for classification due to the adverse impact of the distant local features, the "peakiness" effect and the global bandwidth (see Sect. 3.3.1 for details). Such disadvantages undermine the power of the generative model. As a result, the performance of the traditional soft-assignment coding scheme is suboptimal when combined with the discriminative model.

In this chapter, we propose an improved soft-assignment coding (ISAC) method, which re-interprets the soft-assignment coding in terms of kernel density estimation. Based on the interpretation, we further propose two improvements, namely the thresholding normalized visual word plausibility (TNVWP), and the power

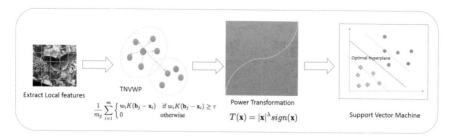

Fig. 3.1 The system architecture of the improved soft-assignment coding (ISAC) method. The local features are first extracted from the image. An image representation vector is then derived by the thresholding normalized visual word plausibility (TNVWP) method. The image representation vector is furthermore transformed by means of the power transformation (PT) method. And finally a support vector machine is utilized for image classification

transformation (PT) method. The system architecture of the improved soft-assignment coding method is shown in Fig. 3.1. The TNVWP method estimates the distribution of each visual word given an image by using only the local features with larger normalized visual word plausibility than a threshold. The reason is that the distant local features might have adverse impact on the probability density estimation. The power transformation (PT) then brings two benefits. First, the TNVWP often derives sparse image vectors especially when the codebook size increases or in the case where the vectors are extracted from sub-regions (spatial pyramid). It will cause the "peakiness" effect when using the dot-product operation in the linear support vector machine. The power transformation is able to cancel out the "peakiness" effect by densifying the image vectors [14]. Second, the power transformation is able to downplay the influence of the local features (usually bursty visual elements) that happen frequently within a given image [14].

As a result, the ISAC method not only inherits the computational efficiency of the traditional soft-assignment coding scheme, but also boosts the performance to be comparable to other popular feature coding methods such as the sparse coding method and the locality-constrained linear coding [9] method on four representative datasets: the UIUC sports event dataset [19], the scene 15 dataset [4], the Caltech 101 dataset [10], and the Caltech 256 dataset [11].

3.2 Related Work

The literature of image classification is vast, thus this chapter only focuses on feature coding methods. Current feature coding methods are mainly developed from two views: the reconstruction point of view and the density estimation point of view.

The first point of view is the reconstruction view, which attempts to model the local feature as a linear combination of codebook visual words with some smoothing constraints or non-smoothing sparsity constraints and then encode the local feature as the vector of coefficients. Many methods [8, 9, 20–22] are proposed based on the reconstruction point of view. In [8], Yang et al. propose to use the sparse coding method to learn codebook and the vector of coefficients to encode the local features. In [9], the locality-constrained linear coding method is proposed which considers the locality information in the feature coding process. In [20], Gao et al. further propose the Laplacian sparse coding, which considers preserving both the similarity and locality information among local features.

Another point of view is the density estimation view, which aims at providing a reliable probability density estimation either over the visual words [4–6] or over the local features [13, 14]. The bag of visual words method [4] counts the frequency of the local features over the visual words and represents the image as a histogram. A more robust alternative to histograms is the kernel density estimation, which induces the soft-assignment coding method [5, 6]. Recently, the Fisher vector method [13, 14] is proposed and it estimates the density over the local features by using the Gaussian mixture model. Then the Fisher kernel based support vector machine is used for classification (the discriminative model).

The topic of this chapter is to improve the soft-assignment coding, and it is worth noting some prior research work. Liu et al. [7] revisit the soft-assignment coding and propose to localize the soft-assignment coding by using the k-nearest visual words motivated by [9]. Our proposed method applies the thresholding normalized visual word plausibility and the power transformation to further improve the soft-assignment coding method.

3.3 The Improved Soft-Assignment Coding

Soft-assignment Coding (SAC) is proposed to model the visual word ambiguity explicitly [5]. This section re-visits the SAC and argues that the kernel density estimation of SAC is not reliable. Based on such an argument, two improvements, namely the thresholding normalized visual word plausibility (TNVWP) and the power transformation, are proposed to establish the relation between SAC and the recently proposed VLAD coding method from the perspective of kernel density estimation.

3.3.1 Revisiting the Soft-Assignment Coding

Soft-assignment Coding resolves the mismatch of the hard assignment coding with the nature of continuous image features. Given the codebook $\mathbf{B} = [\mathbf{b}_1, \mathbf{b}_2, \ldots, \mathbf{b}_k] \in \mathbb{R}^{n \times k}$, for each local feature $\mathbf{x}_i \in \mathbb{R}^n$, $(i = 1, 2, \ldots, m)$, e.g. the SIFT feature, the hard assignment coding $\mathbf{c}_i = [c_{i1}, c_{i2}, \ldots, c_{ik}]^t \in \mathbb{R}^k$ attempts to activate only one non-zero coding coefficient, which corresponds to the nearest visual word in the codebook $\mathbf{B}$ as follows:

$$c_{ij} = \begin{cases} 1 & \text{if } j = \arg\min ||\mathbf{x}_i - \mathbf{b}_j||^2 \\ 0 & \text{otherwise} \end{cases} \tag{3.1}$$

Then the image is represented by pooling over all the local features as $\frac{1}{m}\sum_{i=1}^{m}\mathbf{c}_i$. The pooling of the hard-assignment coding $\mathbf{c}_i$ represents the image as a histogram of visual word frequencies that describes the probability density over the visual words.

However, this kind of 0–1 hard assignment violates the ambiguous nature of local features. As a robust alternative, the soft-assignment coding of $\mathbf{c}_i$ is then proposed as follows:

$$c_{ij} = \frac{exp(-||\mathbf{x}_i - \mathbf{b}_j||^2/2\sigma^2)}{\sum_{j=1}^{k} exp(-||\mathbf{x}_i - \mathbf{b}_j||^2/2\sigma^2)} \tag{3.2}$$

where σ is the smoothing parameter that controls the degree of smoothness of the assignment and $exp(\cdot)$ is the exponential function. Then the image is represented by pooling over all the local features as $\frac{1}{m}\sum_{i=1}^{m}\mathbf{c}_i$. The pooling of the soft-assignment coding replaces the histogram by using the kernel density estimation to estimate the probability density. Usually, the Gaussian kernel is used to describe the relation between the local features and the visual words.

As a matter of fact, at the very beginning, the term "soft-assignment coding" is called the "visual word/codeword uncertainty" in [5], which is essentially a weighted version of kernel density estimation. Given the Gaussian kernel $K(\mathbf{x}) = \frac{1}{\sqrt{2\pi}\sigma}exp(-\frac{\mathbf{x}^2}{2\sigma^2})$, let's first define the weight as follows:

$$w_i = \frac{1}{\sum_{j=1}^{k} K(\mathbf{x}_i - \mathbf{b}_j)} \tag{3.3}$$

The weight is a normalization operator that indicates how much $\mathbf{x}_i$ will contribute to the image representation over all the visual words.

Then the final estimation over each visual word $\mathbf{b}_j(j = 1, 2, \ldots, k)$ will be:

$$\frac{1}{m}\sum_{i=1}^{m} w_i K(\mathbf{b}_j - \mathbf{x}_i) \tag{3.4}$$

where σ in the Gaussian kernel plays the role of a bandwidth in the kernel density estimation.

According to the estimation in Eq. 3.4, we point out the following drawbacks of the soft-assignment coding method that lead to its inferior performance to other popular methods.

- First, the soft-assignment coding considers the visual word uncertainty but does not take the visual word plausibility [5] into account. Because all the m local features are used to compute the distribution over each visual word $\mathbf{b}_j(j = 1, 2, \ldots, k)$, which keeps the distant local features that may have adverse impact. Thus, it is beneficial to consider the visual word plausibility, which selects the local features that best fit the visual word.

- Second, the image vectors derived by the soft-assignment coding are often sparse when the codebook size increases or in the case where the vectors are extracted from sub-regions (spatial pyramid). It will cause the "peakiness" effect, which may harm the performance when using the dot-product operation in the linear support vector machine.
- Third, the soft-assignment coding uses a global bandwidth σ. However, the visual words often have different amounts of smoothing at different locations (the shape of each cluster cannot be fit by the sole bandwidth [7, 23]). Thus, a careful tuning of the bandwidth is necessary.

We next briefly introduce the Fisher vector [13] and the VLAD method [13]. We then present the thresholding normalized visual word plausibility (TNVWP) and the power transformation to address these drawbacks of the soft-assignment coding in order to improve its performance. We finally present our findings about a theoretical relation between our method and the VLAD method.

3.3.2 Introduction to Fisher Vector and VLAD Method

We first briefly review the Fisher vector method [13], which has been widely applied for visual recognition problems such as face recognition [24] and object recognition [13]. Specifically, let $\mathbf{X} = \{\mathbf{d}_t, t = 1, 2, \ldots, T\}$ be the set of T local descriptors (e.g. SIFT descriptors) extracted from the image. Let μ_λ be the probability density function of $\mathbf{X}$ with a set of parameters λ, then the Fisher kernel [25] is defined as follows:

$$K(\mathbf{X}, \mathbf{Y}) = (\mathbf{G}_\lambda^{\mathbf{X}})^T \mathbf{F}_\lambda^{-1} \mathbf{G}_\lambda^{\mathbf{Y}} \tag{3.5}$$

where $\mathbf{G}_\lambda^{\mathbf{X}} = \frac{1}{T} \nabla_\lambda \log[\mu_\lambda(\mathbf{X})]$, which is the gradient vector of the log-likelihood that describes the contribution of the parameters to the generation process. And $\mathbf{F}_\lambda$ is the Fisher information matrix of μ_λ.

Essentially, the Fisher vector is derived from the explicit decomposition of the Fisher kernel according to the fact that the symmetric and positive definite Fisher information matrix $\mathbf{F}_\lambda$ has a Cholesky decomposition as follows:

$$\mathbf{F}_\lambda^{-1} = \mathbf{L}_\lambda^T \mathbf{L}_\lambda \tag{3.6}$$

Therefore, the Fisher kernel $K(\mathbf{X}, \mathbf{Y})$ can be written as a dot product between two vectors $\mathbf{L}_\lambda \mathbf{G}_\lambda^{\mathbf{X}}$ and $\mathbf{L}_\lambda \mathbf{G}_\lambda^{\mathbf{Y}}$ which are defined as the **Fisher vectors** of $\mathbf{X}$ and $\mathbf{Y}$ respectively. A diagonal closed-form approximation of $\mathbf{F}_\lambda$ [13] is often used where $\mathbf{L}_\lambda$ is just a whitening transformation for each dimension of $\mathbf{G}_\lambda^{\mathbf{X}}$ and $\mathbf{G}_\lambda^{\mathbf{Y}}$. The dimensionality of Fisher vector depends only on the number of parameters in the parameter set λ.

More specifically, if we choose the probability density function μ_λ to be a Gaussian mixture model (GMM) with the mixture weight w_i, mean vector μ_i and variance matrix (diagonal assumption) σ_i as follows:

$$\mu_\lambda(x) = \sum_{i=1}^{K} w_i g_i(x : \mu_i, \sigma_i) \tag{3.7}$$

Now the parameter set $\lambda = \{w_i, \mu_i, \sigma_i, i = 1, 2, \ldots, K\}$. The GMM can be trained off-line by maximum likelihood estimation using the local feature descriptors that are extracted from the images. The diagonal closed-form approximation of the Fisher information matrix $\mathbf{F}_\lambda$ [13] can be derived from the GMM so that $\mathbf{L}_\lambda$ is just a whitening transformation for each dimension of $\mathbf{G}_\lambda^X$ and $\mathbf{G}_\lambda^Y$, which is proportional to σ_i^{-1}. Then the Fisher vector can be defined as follows:

$$f = [\frac{1}{T\sqrt{w_i}} \sum_{t=1}^{T} \gamma_t(i)\sigma_i^{-1}(x_t - \mu_i)](i = 1, 2, \ldots, K) \tag{3.8}$$

where $\gamma_t(i) = \frac{w_i g_i(x_t)}{\sum_{j=1}^{K} w_j g_j(x_t)}$. As a result, the final dimensionality of the Fisher vector f is $K \times d$, where d is the dimension of the local feature descriptor, e.g. 64 for SIFT feature. If the variance is considered, the gradient score of variance is also concatenated into the final vector, which leads to a dimensionality of $2K \times d$.

The VLAD representation is a non probabilistic approximation of the Fisher vector, where the k-means algorithm is applied to derive the centroids for replacing the GMM as follows:

$$f = [\sum_{x_t}(x_t - \mu_i)](i = 1, 2, \ldots, K) \tag{3.9}$$

where x_t belongs to the cluster with the centroid μ_i.

As shown in Fig. 3.2, the key difference among the conventional bag-of-feature framework (the left image), the Fisher vector and VLAD framework (the middle image), and our proposed method (right image) is the way of feature coding.

3.3.3 The Thresholding Normalized Visual Word Plausibility

Soft-assignment coding only considers the visual word uncertainty (VWU) but does not take the visual word plausibility (VWP) [5] into account. The VWU and VWP are contradictory since the VWU indicates that a local feature might contribute to the probability distribution of more than one visual words. While, the VWP signifies that a local feature may not be close enough to contribute to the density estimation of the relevant visual word, which means one local feature only contributes to the

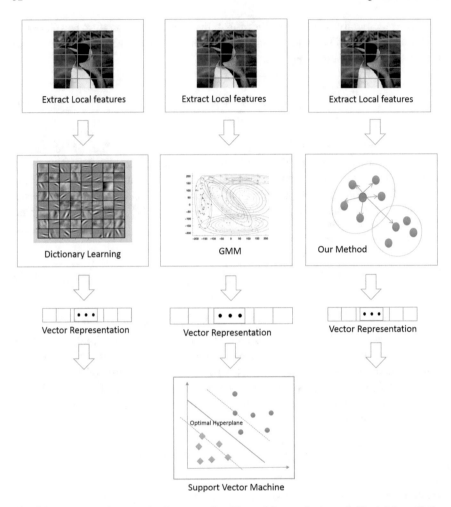

Fig. 3.2 The comparison among the conventional bag-of-feature framework (the *left* image), the Fisher vector and VLAD framework (the *middle* image), and our proposed method (*right* image)

visual word in the same cluster. This can be seen from the definition of visual word plausibility in Eq. 3.10 that the visual word plausibility only uses the local features in the same cluster as the visual word for density estimation while discards all the other distant features.

$$\frac{1}{m}\sum_{i=1}^{m} \begin{cases} K(\mathbf{b}_j - \mathbf{x}_i) & \text{if } j = \arg\min_{1 \le v \le k} ||\mathbf{x}_i - \mathbf{b}_v||^2 \\ 0 & \text{otherwise} \end{cases} \tag{3.10}$$

However, the work in [5] finds that visual word plausibility alone can only achieve disappointing inferior results comparing to the soft-assignment coding (visual word uncertainty).

Thus, to exploit the advantages of both, we propose the thresholding normalized visual word plausibility (TNVWP), which trades off both the visual word uncertainty and the visual word plausibility. The TNVWP consists of two steps: a normalization step and a visual word plausibility thresholding step.

(1) Mathematically, first, we compute $w_i K (\mathbf{b}_j - \mathbf{x}_i)$ for all the local features $\mathbf{x}_i$ ($i = 1, 2, \ldots, m$) and all the visual words $\mathbf{b}_j$ ($j = 1, 2, \ldots, k$).

(2) Then for a given image I, we select local features that are close enough to the relevant visual word to estimate the density estimation over the relevant visual word by using the following equations:

$$\frac{1}{m_j} \sum_{i=1}^{m} \begin{cases} w_i K (\mathbf{b}_j - \mathbf{x}_i) & \text{if } w_i K (\mathbf{b}_j - \mathbf{x}_i) \geq \tau \\ 0 & \text{otherwise} \end{cases} \tag{3.11}$$

where τ works as the threshold, m_j is the number of local features $\mathbf{x}_i$ that satisfy the condition $w_i K (\mathbf{b}_j - \mathbf{x}_i) \geq \tau$.

On the one hand, the TNVWP keeps the property of visual word uncertainty that each local feature might contribute to more than one visual word if the visual word is close enough. On the other hand, the TNVWP assigns larger weight to the local features that are close to the visual word. The parameter τ is used to trade off the two aspects of the TNVWP.

3.3.4 The Power Transformation

The final image representations derived by the TNVWP are often sparse especially when the codebook size increases or in the case where the vectors are extracted from sub-regions (spatial pyramid). It will cause the "peakiness" effect when using the dot-product operation in the linear support vector machine because some values in the sparse vectors might vanish in the final value of the dot-product operation, which may hinder the classification performance. Moreover, the local features which happen frequently within a given image are more likely to be the visual bursty elements which are not distinctive features. These noisy local features will result in larger values in the final sparse image vectors and dominate the final value of the dot-product operation.

Thus, we propose to apply the power transformation defined as follows:

$$T(\mathbf{x}) = |\mathbf{x}|^\lambda sign(\mathbf{x}) \tag{3.12}$$

where all the operations are element-wise and $sign(\mathbf{x})$ is the sign vector indicating the sign of each element of $\mathbf{x}$.

The power transformation has been widely used in such cases [13, 14] and it is able to cancel out the "peakiness" effect by densifying the image vectors. Second, it is able to decrease the influence of local features (usually bursty visual elements) which happen frequently within a given image [14]. The L_2 normalization always follows the power transformation.

3.3.5 Relation to VLAD Method

In this section, we aim at establishing a relation between the TNVWP with the VLAD method. Assuming that C_j is the set of local features that satisfy the condition $w_i K(\mathbf{b}_j - \mathbf{x}_i) \geq \tau$, then the final image representation $\mathbf{v}$ will be as follows:

$$[\sum_{x_i \in C_1} \frac{w_i}{m_1} K(\mathbf{b}_1 - \mathbf{x}_i), \ldots, \sum_{x_i \in C_k} \frac{w_i}{m_k} K(\mathbf{b}_k - \mathbf{x}_i)]^t \qquad (3.13)$$

By using the simple Taylor expansion, $K(\mathbf{b}_j - \mathbf{x}_i) \approx K(0) + \nabla K^t(0)(\mathbf{b}_j - \mathbf{x}_i)$, and dropping the high order terms, we have

$$\mathbf{v} = \mathbf{w}_0 + \mathbf{W}[\sum_{x_i \in C_j} \frac{w_i}{m_j} (\mathbf{b}_j - \mathbf{x}_i)]_{(n*k) \times 1} \qquad (3.14)$$

where $\mathbf{w}_0 = [\sum_{x_i \in C_1} \frac{w_i}{m_1} K(0), \ldots, \sum_{x_i \in C_k} \frac{w_i}{m_k} K(0)]^t$, which is a constant vector, and $\mathbf{W} = [\nabla K^t(0)]_{k \times (n*k)}$, which is a matrix of size $k \times (n * k)$ that the constant vector $\nabla K^t(0) \in \mathbb{R}^{1 \times n}$ is organized in the diagonal form.

It can be revealed that if the threshold is adaptively selected so that C_j is the cluster obtained by the k-means method, then $[\sum_{x_i \in C_j} \frac{w_i}{m_j} (\mathbf{b}_j - \mathbf{x}_i)]_{(n*k) \times 1}$ is a weighted version of the VLAD method [13]. In [13], VLAD is usually followed by the power transformation and the L_2 normalization, which is proven to be effective. Thus, it is reasonable to apply the power transformation and the L_2 normalization to improve the soft-assignment coding as stated in the above sections.

3.4 Experiments

In this section, the proposed method is assessed on four popular datasets: the UIUC sports event dataset [19], the scene 15 dataset [4], the Caltech 101 dataset [10] and the Caltech 256 dataset [11] with the intention to show the feasibility of the proposed method. In addition, we conduct experiments to show the impact of the model parameters for an in-depth analysis of the proposed method.

It has been empirically shown that the local feature plays an important role in the final performance of image classification. For a fair comparison, we use the widely adopted dense SIFT feature [4]. Specifically, the dense SIFT feature is extracted from only one single scale with the step size as 8 pixels and the patch size as 16 pixels for the UIUC sports event dataset and the scene 15 dataset. With respect to the Caltech 101 dataset and the Caltech 256 dataset, we follow the setting in [9] that the dense SIFT feature is extracted with the step size as 8 pixels under three scales, 16×16, 25×25 and 31×31 respectively. We also rescale the maximum side (width or height) of the image to 300 pixels with the preserved aspect ratio for all the datasets except the UIUC sports event dataset, which is rescaled to 400 pixels due to the high resolution of it [7]. For the codebook size, 1024 is used for the UIUC sports event dataset, the scene 15 dataset and 2048 is used for Caltech 101 dataset and the Caltech 256 dataset. The spatial pyramid representation with 3 levels 1×1, 2×2 and 4×4 is used for all the experiments. As for classification, the liblinear package [26] for fast linear kernel based support vector machine is used for all the experiments.

3.4.1   The UIUC Sports Event Dataset

The UIUC sports event dataset [19] consists of 8 sports event categories with a total number of 1574 images that are collected from the Internet. The number of images in each sport varies from 137 to 250. Example images are shown in Fig. 3.3.

Following the common experimental setting [19], we randomly select 70 images for training and 60 images for testing in each event class. We report the average accuracy of the proposed method on 10 random training/testing splits in Table 3.1. The threshold τ is set as 0.05 and the bandwidth $\frac{1}{2\sigma^2}$ is set as 20. The parameter $\alpha = 0.5$ is selected for the power transformation. For the liblinear package, the parameters are set as $c = 16$. As shown in Table 3.1, we compare the proposed

Fig. 3.3 Example images of the UIUC Sports Event Dataset

Table 3.1 Comparisons between the proposed method and the other popular methods on the UIUC sports event dataset. PT is the acronym for power transformation

Methods	Accuracy %
SIFT + GGM [19]	73.4
OB [27]	76.3
CA-TM [28]	78.0
SAC [5]	82.04 ± 2.37
ScSPM [8]	82.74 ± 1.46
LLC [9]	81.41 ± 1.84
LSAC [7]	82.29 ± 1.84
LSC [20]	85.18 ± 0.46
HMP [21]	85.7 ± 1.3
The ISAC method without PT	**83.38** ± 1.57
The ISAC method with PT	**86.00** ± 1.19

method with other popular methods such as the "SIFT + GGM" (SIFT + generative graphical model) method [19], the object bank (OB) method [27], the context aware topic model (CA-TM) method [28], the traditional soft-assignment coding (SAC) method [5], the "ScSPM" method [8], the locality-constrained linear coding (LLC) method [9], the localized soft-assignment coding (LSAC) method [7], the Laplacian sparse coding (LSC) method [20] and the hierarchical match pursuit (HMP) method [21]. The proposed method is able to outperform or be comparable to these popular methods but with a much faster speed. The results also prove the effectiveness of the power transformation that the ISAC method with the power transformation improves upon the one without power transformation by more than 2 percent. Please note that the results are achieved under the condition that the SIFT feature, the size of the visual word, the scale ratio of the image are all the same for fair comparison.

3.4.2 The Scene 15 Dataset

The scene 15 dataset [4] contains totally 4485 images from 15 scene categories, each with the number of images ranging from 200 to 400. Example images are shown in Fig. 3.4.

Following the experimental protocol defined in [4, 8], we randomly select 100 images per class for training and the remaining for testing for 10 iterations. The threshold τ is set as 0.01 and the bandwidth $\frac{1}{2\sigma^2}$ is set as 40. The parameter $\alpha = 0.3$ is selected for the power transformation. For the liblinear package, the parameters are set as $c = 16$, $B = 4$. We choose the following methods for comparison: spatial pyramid matching with non-linear kernel support vector machine (KSPM) [4], the traditional soft-assignment coding (SAC) method [6], the locally constrained linear

Fig. 3.4 Example images of the Scene 15 Dataset

Table 3.2 Comparisons between the proposed method and the other popular methods on the scene 15 dataset

Methods	Accuracy %
KSPM [4]	81.40 ± 0.50
ScSPM [8]	80.28 ± 0.93
LLC [9]	$80.57 \pm -$
SAC [6]	76.67 ± 0.93
The ISAC method without PT	**79.55** ± 0.46
The ISAC method with PT	**82.79** ± 0.47

coding (LLC) [9] and the sparse coding method (ScSPM) [8]. All the compared methods use non-linear or linear kernel based support vector machine for classification. The results are shown in Table 3.2, it is seen that the proposed method is able to achieve comparable results to the other popular methods. It is worth noting that the ISAC method with the power transformation improves upon the one without power transformation by 3%, which demonstrates the effectiveness of the proposed power transformation method.

3.4.3 The Caltech 101 Dataset

The Caltech 101 dataset [10] holds 9144 images divided into 101 object classes and a clutter class. The number of images per category varies from 31 to 800. Example images are shown in Fig. 3.5.

To be consistent with the previous work [9], we partition the whole dataset randomly into 15, 20, 25, 30 training images per class and no more than 50 testing images per class, and measure the performance using the average accuracy over 102 classes. We train a codebook with 2048 visual words.

Fig. 3.5 Example images of the Caltech 101 Dataset

Table 3.3 Comparisons between the proposed method and the other popular methods on the Caltech 101 dataset

Training images	15	20	25	30
SVM-KNN [29]	59.10	62.00	–	66.20
SPM [4]	56.40	–	–	64.60
Griffin [11]	59.00	63.30	65.80	67.60
Jain [30]	61.00	–	–	69.10
Boiman [31]	65.00	–	–	70.40
Gemert [6]	–	–	–	64.16
SAC [2]	–	–	–	71.5
ScSPM [8]	67.00	–	–	73.20
LLC [9]	65.43	67.74	70.16	73.44
Our method without PT	63.98	**67.88**	**70.16**	73.20
Our method with PT	**67.92**	**71.29**	**73.53**	**75.39**

The threshold τ is set as 0.02 and the bandwidth $\frac{1}{2\sigma^2}$ is set as 30. The parameter $\alpha = 0.5$ is selected for the power transformation. For the liblinear package, the parameters are set as $c = 1$. The results that are shown in Table 3.3, demonstrate the proposed method is able to outperform the other popular methods in all the training image size. Please note that the experiment settings, such as the dense SIFT feature used, the size of codebook, the spatial pyramid configurations, are all the same to the ones in [9] for fair comparison. Another issue is the evaluation metric used, the result in Table 3.3 is the average accuracy over all the classes.

Please note that some papers [7] also report the results of classification accuracy. In terms of classification accuracy, the proposed method can achieve 77.42 ± 0.63, which is better than 76.48 ± 0.71 in [7].

3.4.4   The Caltech 256 Dataset

The Caltech 256 dataset [11] contains 30,607 images divided into 256 object categories and a clutter class. The challenge comes from the high intra-class variability and object location variability. Each class contains at least 80 images. Some example images are shown in Fig. 3.6.

We follow the common experimental settings [9] that 15, 30, 45, 60 images per category are selected randomly for training and no more than 25 images for testing. We also train a codebook with 2048 visual words. The threshold τ is set to 0.02 and the bandwidth $\frac{1}{2\sigma^2}$ is set to 20. The parameter $\alpha = 0.5$ is selected for the power transformation. For the liblinear package, the parameters are set as $c = 1.5$. The results that are shown in Table 3.4, demonstrate the proposed method is able to achieve comparable image classification performance to other popular methods.

Fig. 3.6   Example images of the Caltech 256 Dataset

Table 3.4 Comparisons between the proposed method and some other popular methods on the Caltech 256 dataset

Training images	15	30	45	60
Griffin [11]	28.30	34.10	–	–
SAC [6]	–	27.17	–	–
ScSPM [8]	27.73	34.02	37.46	40.14
LLC [9]	34.36	41.19	45.31	47.68
The proposed method	32.67	39.11	42.35	44.69

3.4.5   In-depth Analysis

In this section, we provide more comprehensive analysis of the proposed method. The performance of the proposed method is evaluated when two key parameters change, namely the change of the bandwidth parameter and change of the threshold value on the Caltech 101 dataset.

As shown in Fig. 3.7, we evaluate the performance when the value of β changes ($\beta = \frac{1}{2\sigma^2}$), which is the inverse of twice of the square of the bandwidth parameter σ. It is seen that large values of β (small value of σ, hence small bandwidth) often achieve better results.

The assessment on the value of the threshold is also conducted, as shown in Fig. 3.8, the value of τ needs to be carefully selected.

Fig. 3.7 The performance measured when the value of β changes ($\beta = \frac{1}{2\sigma^2}$) changes on the Caltech 101 dataset

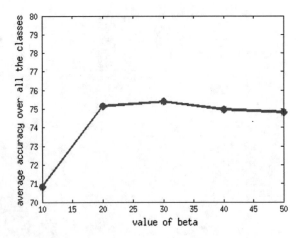

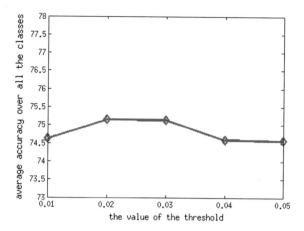

Fig. 3.8 The performance measured when the value of threshold τ changes on the Caltech 101 dataset

3.5   Conclusion

This chapter presents the improved soft-assignment coding (ISAC) method by introducing two enhancements, namely, the thresholding normalized visual word plausibility (TNVWP) and the power transformation method. Such improvements are then shown to be theoretically reasonable and establish the relation between the proposed ISAC method and the VLAD coding method. The proposed ISAC method is further discussed and evaluated on four representative datasets: the UIUC sports event dataset, the scene 15 dataset, the Caltech 101 dataset, and the Caltech 256 dataset. The proposed ISAC method achieves competitive results to and even better results than some popular image classification methods without sacrificing much computational efficiency.

References

1. Huang, Y., Wu, Z., Wang, L., Tan, T.: Feature coding in image classification: A comprehensive study. IEEE Trans. Pattern Anal. Mach. Intell. pp. 493–506 (2014)
2. Boureau, Y., Bach, F., LeCun, Y., Ponce, J.: Learning mid-level features for recognition. In: Proceedings of the International Conference on Computer Vision and Pattern Recognition (CVPR'10) (2010)
3. Boureau, Y., Ponce, J., LeCun, Y.: A theoretical analysis of feature pooling in vision algorithms. In: Proceedings of the International Conference on Machine learning (ICML' 10) (2010)
4. Lazebnik, S., Schmid, C., Ponce, J.: Beyond bags of features: Spatial pyramid matching for recognizing natural scene categories. In: IEEE Conference on Computer Vision and Pattern Recognition, pp. 2169–2178 (2006)
5. van Gemert, J.C., Veenman, C.J., Smeulders, A.W.M., Geusebroek, J.M.: Visual word ambiguity. IEEE Trans. Pattern Anal. Mach. Intell. **32**(7), 1271–1283 (2010)
6. Gemert, J.C., Geusebroek, J.M., Veenman, C.J., Smeulders, A.W.: Kernel codebooks for scene categorization. In: European Conference on Computer Vision, pp. 696–709 (2008)

7. Liu, L., Wang, L., Liu, X.: In defense of soft-assignment coding. In: Proceedings of the 2011 International Conference on Computer Vision, pp. 2486–2493 (2011)
8. Yang, J., Yu, K., Gong, Y., Huang, T.S.: Linear spatial pyramid matching using sparse coding for image classification. In: IEEE Conference on Computer Vision and Pattern Recognition, pp. 1794–1801 (2009)
9. Wang, J., Yang, J., Yu, K., Lv, F., Huang, T.S., Gong, Y.: Locality-constrained linear coding for image classification. In: IEEE Conference on Computer Vision and Pattern Recognition, pp. 3360–3367 (2010)
10. Fei-Fei, L., Fergus, R., Perona, P.: Learning generative visual models from few training examples. In: IEEE Proceedings of the Workshop on Generative-Model Based Vision, CVPR (2004)
11. Griffin, G., Holub, A., Perona, P.: Caltech-256 object category dataset. Technical report 7694. California Institute of Technology (2007). http://authors.library.caltech.edu/7694
12. Jégou, H., Douze, M., Schmid, C., Pérez, P.: Aggregating local descriptors into a compact image representation. In: IEEE Conference on Computer Vision and Pattern Recognition, pp. 3304–3311 (2010)
13. Jegou, H., Perronnin, F., Douze, M., Sanchez, J., Perez, P., Schmid, C.: Aggregating local image descriptors into compact codes. IEEE Trans. Pattern Anal. Mach. Intell. **34**(9), 1704–1716 (2012)
14. Snchez, J., Perronnin, F., Mensink, T., Verbeek, J.J.: Image classification with the fisher vector: Theory and practice. Int. J. Comput. Vis. **105**(3), 222–245 (2013)
15. Arandjelović, R., Zisserman, A.: All about VLAD. In: IEEE Conference on Computer Vision and Pattern Recognition (2013)
16. Bosch, A., Zisserman, A., Munoz, X.: Scene classification using a hybrid generative/discriminative approach. IEEE Trans. Pattern Anal. Mach. Intell. **30**(4) (2008)
17. Perina, A., Cristani, M., Castellani, U., Murino, V., Jojic, N.: Free energy score spaces: Using generative information in discriminative classifiers. IEEE Trans. Pattern Anal. Mach. Intell. **34**(7), 1249–1262 (2012)
18. Jebara, T., Kondor, R., Howard, A.: Probability product kernels. J. Mach. Learn. Res. **5**, 819–844 (2004)
19. jia Li, L., fei Li, F.: What, where and who? Classifying event by scene and object recognition. In: IEEE International Conference on Computer Vision (2007)
20. Gao, S., Tsang, I.W.H., Chia, L.T.: Laplacian sparse coding, hypergraph laplacian sparse coding, and applications. IEEE Trans. Pattern Anal. Mach. Intell. **35**(1), 92–104 (2013)
21. Bo, L., Ren, X., Fox, D.: Hierarchical matching pursuit for image classification: Architecture and fast algorithms. In: Advances in Neural Information Processing Systems (2011)
22. Zhou, X., Yu, K., Zhang, T., Huang, T.S.: Image classification using super-vector coding of local image descriptors. In: Proceedings of the 11th European Conference on Computer Vision: Part V, pp. 141–154 (2010)
23. Wand, M.P., Marron, J.S., Ruppert, D.: Transformations in density estimation. J. Am. Stat. Assoc. **86**(414), 343–353 (1991)
24. Simonyan, K., Parkhi, O.M., Vedaldi, A., Zisserman, A.: Fisher vector faces in the wild. In: British Machine Vision Conference (BMVC) (2013)
25. Jaakkola, T., Haussler, D.: Exploiting generative models in discriminative classifiers. In: Advances in Neural Information Processing Systems, pp. 487–493 (1998)
26. Fan, R.E., Chang, K.W., Hsieh, C.J., Wang, X.R., Lin, C.J.: Liblinear: a library for large linear classification. J. Mach. Learn. Res. **9**, 1871–1874 (2008)
27. Li, L.J., Su, H., Xing, E.P., Li, F.F.: Object bank: A high-level image representation for scene classification and semantic feature sparsification. In: Advances in Neural Information Processing Systems, pp. 1378–1386 (2010)
28. Niu, Z., Hua, G., Gao, X., Tian, Q.: Context aware topic model for scene recognition. In: IEEE Conference on Computer Vision and Pattern Recognition, pp. 2743–2750 (2012)

29. Zhang, H., Berg, A.C., Maire, M., Malik, J.: Svm-knn: Discriminative nearest neighbor classification for visual category recognition. In: Proceedings of the 2006 IEEE Computer Society Conference on Computer Vision and Pattern Recognition, pp. 2126–2136 (2006)
30. Jain, P., Kulis, B., Grauman, K.: Fast image search for learned metrics. In: IEEE Conference on Computer Vision and Pattern Recognition, CVPR 2008, pp. 1–8. IEEE (2008)
31. Boiman, O., Shechtman, E., Irani, M.: In defense of nearest-neighbor based image classification. In: IEEE Conference on Computer Vision and Pattern Recognition, CVPR 2008, pp. 1–8. IEEE (2008)

Chapter 4
Inheritable Color Space (InCS) and Generalized InCS Framework with Applications to Kinship Verification

Qingfeng Liu and Chengjun Liu

Abstract This chapter presents a novel inheritable color space (InCS) and a generalized InCS (GInCS) framework for kinship verification. Unlike conventional color spaces, the proposed InCS is automatically derived by balancing the criterion of minimizing the distance between kinship images and the criterion of maximizing the distance between non-kinship images based on a new color similarity measure (CSM). Two important properties of the InCS, namely, the decorrelation property and the robustness to illumination variations property, are further presented through both theoretical and practical analyses. To utilize other inheritable features, a generalized InCS framework is then presented to extend the InCS from the pixel level to the feature level for improving the verification performance as well as the robustness to illumination variations. Experimental results on four representative datasets, the KinFaceW-I dataset, the KinFaceW-II dataset, the UB KinFace dataset, and the Cornell KinFace dataset, show the effectiveness of the proposed method.

4.1 Introduction

Kinship verification from facial images, which is an emerging research area in computer vision, has gained increasing attention in recent years [1–6]. Pioneer works in anthropology [7, 8] believe that there are some genetic related features which are inherited by children from their parents that can be used to determine the kinship relations. In addition to shape and texture, color is another important inheritable feature. Our observation shows that color information, cooperating with shape and texture, usually leads to better performance for kinship verification than those methods without color (the cross-ethnicity parents and children are not considered in this chapter since such cases are minority).

Q. Liu (✉) · C. Liu
New Jersey Institute of Technology, Newark, NJ 07102, USA
e-mail: ql69@njit.edu

C. Liu
e-mail: cliu@njit.edu

© Springer International Publishing AG 2017
C. Liu (ed.), *Recent Advances in Intelligent Image Search and Video Retrieval*,
Intelligent Systems Reference Library 121, DOI 10.1007/978-3-319-52081-0_4

As a complementary feature to shape and texture, color have proved its effectiveness for face recognition [9–13]. Conventional color spaces, e.g. RGB, have shown their superiority for face recognition [11, 14–16]. Recently many color methods [17–22] are proposed for object recognition, object detection and action recognition. These methods, which focus on illumination invariant properties of color, are not specifically designed for capturing the inheritable information between kinship images for recognizing the kinship relations.

This chapter thus presents a novel inheritable color space and a generalized InCS framework for kinship verification. Specifically, a new color similarity measure (CSM), which represents the accumulation of the similarity measures of the corresponding color components between two images, is first defined. A novel inheritable color space (InCS) is then automatically derived by balancing between the criterion of minimizing the distance between the kinship images and the criterion of maximizing the distance between the non-kinship images. Some properties of InCS are further discussed. Theoretical analysis shows that the proposed InCS possesses the decorrelation property, which decorrelates the color components of the color difference matrix between two images. Experimental analysis also discovers that the decorrelation property, which is measured in terms of average statistical correlation (ASC), often leads to better performance. The robustness to illumination variations under some mild illumination conditions at both pixel level and feature level is also presented. Finally, to utilize other inheritable features, such as shape and texture, a generalized InCS (GInCS) framework is proposed by extending the InCS from the pixel level to the feature level for improving the performance and the illumination robustness. An example is presented by applying the Fisher vector [23] to extract features from the RGB component images prior to the computation of the component images in the InCS. The system architecture of the proposed InCS is illustrated in Fig. 4.1.

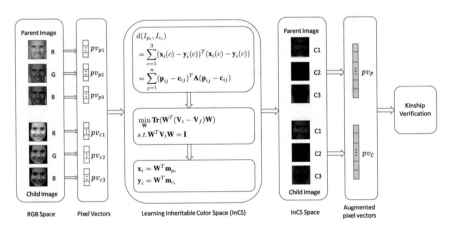

Fig. 4.1 System architecture of the proposed InCS method

Experimental results on four representative kinship verification datasets, namely, the KinFaceW-I dataset, the KinFaceW-II dataset [3], the UB KinFace dataset [2], and the Cornell KinFace dataset [1], show the feasibility of the proposed method.

Our contributions are summarized as follows.

- (i) First, a novel inheritable color space (InCS), which is able to capture the color variations between two images, is presented.
- (ii) Second, our proposed inheritable color space demonstrates the decorrelation property, which leads to better performance. Besides, it also shows robustness to illumination variations under mild illumination conditions at both pixel level and feature level.
- (iii) Third, a generalized InCS (GInCS) framework is presented by extending the InCS from the pixel level to feature level, such as the Fisher vector [23], for improving the performance and illumination invariance.
- (iv) Fourth, a novel color similarity measure (CSM) is proposed to incorporate the information of each color component.

4.2 Related Work

Kinship verification. The pioneer works of kinship analysis originate from anthropology and psychology community. Bressnan et al. [8] evaluated the phenotype matching on facial features and claimed that parents have correlated visual resemblance with their offspring. Studies [7] in anthropology have confirmed that children resemble their parent more than other people and they may resemble a particular parent more at different ages. Later work [1] by Fang et al. showed the feasibility of applying computer vision techniques for kinship verification. Xia et al. [2] proposed a transfer subspace learning based algorithm by using the young parents set as an intermediate set to reduce the significant divergence in the appearance distributions between children and old parents facial images. Lu et al. [3] proposed neighborhood repulsed metric learning (NRML) in which the intraclass samples within a kinship relation are pulled as close as possible and interclass samples are pushed as far as possible for kinship verification. Dehghan et al. [4] proposed to apply the generative and the discriminative gated autoencoders to learn the genetic features and metrics together for kinship verification. Yan et al. [24] proposed a multimetric learning method to combine different complementary feature descriptors for kinship verification and later [5] proposed to learn the discriminative mid-level features by constructing a reference dataset instead of using hand-crafted descriptors. Lu et al. [6] presented the results of various teams on the FG 2015 Kinship Verification in the Wild challenge.

Color space. Color information contributes significantly to the discriminative power of image representation. Conventional color spaces have shown their ability for improving the performance of face recognition [9–16]. For a detail comparison among different color spaces, please refer to [16]. Liu [9] proposed the uncorrelated, independent, and discriminating color spaces for the face recognition problem. Van de Sande et al. [21] showed that color information along with shape features yield excellent results in an image classification system. Khan et al. [19] proposed the use of color attributes as an explicit color representation for object detection. Zhang et al. [22] proposed a new biologically inspired color image descriptor that uses a hierarchical non-linear spatio-chromatic operator yielding spatial and chromatic opponent channels. Khan et al. [25] showed that better results can be obtained for object recognition by explicitly separating the color cue to guide attention by means of a top-down category-specific attention map. Yang et al. [26] proposed a new color model—the $g_1g_2g_3$ model based on the logarithm of the chromacity color space, which preserves the relationship between R, G and B in the model. Rahat Khan et al. [20] clustered color values together based on their discriminative power such that the drop of mutual information of the final representation is minimized.

Metric learning. The motivation of our InCS is closely related to metric learning. Metric learning methods have gained a lot of attention for computer vision and machine learning applications. Earlier work by Xing et al. [27] applied the semi-definite programming to learn a Mahalanobis metric. Goldberger et al. [28] proposed the neighborhood component analysis (NCA) by minimizing the cross validation error of a kNN classifier. Weinberger et al. proposed the large margin nearest neighbor (LMNN) [29] which uses the hinge loss to encourage the related neighbors to be at least one distance unit closer than points from other classes. Davis et al. proposed the information-theoretic metric learning (ITML) [30] by formulating the problem as minimizing the differential relative entropy between two multivariate Gaussian distributions parameterized by the learned metric space and a prior known metric space. Hieu and Li [31] proposed the cosine similarity metric learning (CSML) which utilizes the favorable properties of cosine similarity. Lu et al. [3] proposed the neighborhood repulsed metric learning (NRML) for kinship verification which pays more attention to neighborhood samples.

4.3 A Novel Inheritable Color Space (InCS)

Conventional color spaces have shown their ability for recognition problems [16] by considering the illumination invariant properties. However, they are not deliberately designed to capture the inheritable information between kinship images for recognizing the kinship relations. We therefore present a novel inheritable color space (InCS) for kinship verification by deriving a transformation $\mathbf{W} \in \mathbb{R}^{3 \times 3}$ from the original RGB color space.

In particular, given a pair of images I_{p_i} and I_{c_i} ($i = 1, 2, \ldots, m$) with a size of $h \times w$, these two images can be represented as two matrices $\mathbf{m}_{p_i}$ and $\mathbf{m}_{c_i} \in \mathbb{R}^{3 \times n}$ ($i = 1, 2, \ldots, m$) ($n = h \times w$) respectively, where each row vector is the concatenation of column pixels in each color component (red, green and blue) of the image. Note that in Sect. 4.5, each row vector will be a feature vector (e.g. Fisher vector) for each color component. Formally, $\mathbf{m}_{p_i}$ and $\mathbf{m}_{c_i}$ are defined as follows.

$$\mathbf{m}_{p_i} = [\mathbf{p}_{i1}, \mathbf{p}_{i2}, \ldots, \mathbf{p}_{in}]$$
$$\mathbf{m}_{c_i} = [\mathbf{c}_{i1}, \mathbf{c}_{i2}, \ldots, \mathbf{c}_{in}] \qquad (4.1)$$

where $\mathbf{p}_{ij}$ and $\mathbf{c}_{ij} \in \mathbb{R}^{3 \times 1} (j = 1, 2, \ldots, n)$ are vectors that consists of the color values of three color components at each pixel for two compared images respectively. Then the new representation $\mathbf{x}_i$ and $\mathbf{y}_i \in \mathbb{R}^{3 \times n} (i = 1, 2, \ldots, m)$ in InCS is computed from $\mathbf{m}_{p_i}$ and $\mathbf{m}_{c_i}$ by applying the transformation $\mathbf{W}$ as follows:

$$\mathbf{x}_i = \mathbf{W}^T \mathbf{m}_{p_i} = [\mathbf{x}_{i1}, \mathbf{x}_{i2}, \ldots, \mathbf{x}_{in}]$$
$$\mathbf{y}_i = \mathbf{W}^T \mathbf{m}_{c_i} = [\mathbf{y}_{i1}, \mathbf{y}_{i2}, \ldots, \mathbf{y}_{in}] \qquad (4.2)$$

where $\mathbf{x}_{ij}, \mathbf{y}_{ij} \in \mathbb{R}^{3 \times 1} (j = 1, 2, \ldots, n)$.

Similarity measures [32, 33] are important for the performance. To incorporate the information of all the three color components, we first define a new color similarity measure (CSM). Let $\mathbf{x}_i(c) \in \mathbb{R}^{n \times 1} (c = 1, 2, 3)$ be the c-th color component vector of the InCS in image I_{p_i}, whose values are identical to the c-th row of $\mathbf{x}_i$. And $\mathbf{y}_i(c)$ is defined similarly. Then the CSM has the following mathematical form:

$$
\begin{aligned}
d(I_{p_i}, I_{c_i}) &= \mathbf{Tr}\{(\mathbf{x}_i - \mathbf{y}_i)(\mathbf{x}_i - \mathbf{y}_i)^T\} \\
&= \mathbf{Tr}\{(\mathbf{m}_{p_i} - \mathbf{m}_{c_i})^T \mathbf{A}(\mathbf{m}_{p_i} - \mathbf{m}_{c_i})\} \\
&= \sum_{j=1}^{n} (\mathbf{p}_{ij} - \mathbf{c}_{ij})^T \mathbf{A}(\mathbf{p}_{ij} - \mathbf{c}_{ij})
\end{aligned} \qquad (4.3)
$$

where $\mathbf{Tr}\{\cdot\}$ denotes the trace of a matrix and $\mathbf{A} = \mathbf{W}\mathbf{W}^T \in \mathbb{R}^{3 \times 3}$. Note that since $\mathbf{W}$ is a transformation applied to the RGB color space for deriving a new color space, it is reasonable to assume it is full rank. As a result, $\mathbf{A}$ is a positive semidefinite matrix, which guarantees CSM is a metric—satisfying the non-negativity and the triangle inequality [27].

From Eq. 4.3, the proposed CSM may be interpreted in two ways. On the one hand, the CSM may be interpreted as the accumulation of the distance metrics between corresponding pixel color values of two images as $\sum_{j=1}^{n} (\mathbf{p}_{ij} - \mathbf{c}_{ij})^T \mathbf{A}(\mathbf{p}_{ij} - \mathbf{c}_{ij})$. On the other hand, the CSM may also be interpreted as a summation of the Euclidean similarity measures between the corresponding color components, which is denoted as $\sum_{c=1}^{3} (\mathbf{x}_i(c) - \mathbf{y}_i(c))^T (\mathbf{x}_i(c) - \mathbf{y}_i(c))$.

Then based on such a similarity measure, our inheritable color space (InCS) is derived with the goal of pushing the non-kinship samples as far away as possible while pulling the kinship ones as closely as possible by optimizing the following objective function:

$$\min_{\mathbf{W}} \lambda \sum_{\mathbf{T}} d(I_{p_i}, I_{c_i}) - (1 - \lambda) \sum_{\mathbf{F}} d(I_{p_i}, I_{c_i}) \qquad (4.4)$$

where $\mathbf{T}$ and $\mathbf{F}$ are the sets of kinship image pairs and non-kinship image pairs respectively, and the parameter $\lambda \in (0, 1)$ balances these two terms.

To optimize Eq. 4.4, we further define two matrices as shown in Eq. 4.5, namely the true verification matrix $\mathbf{V}_t \in \mathbb{R}^{3 \times 3}$ that characterizes the color variations among the kinship image pairs, and the false verification matrix $\mathbf{V}_f \in \mathbb{R}^{3 \times 3}$ that captures the color variations among all the non-kinship image pairs.

$$\mathbf{V}_t = \lambda \sum_{\mathbf{T}} \sum_{j=1}^{n} (\mathbf{p}_{ij} - \mathbf{c}_{ij})(\mathbf{p}_{ij} - \mathbf{c}_{ij})^T$$

$$\mathbf{V}_f = (1 - \lambda) \sum_{\mathbf{F}} \sum_{j=1}^{n} (\mathbf{p}_{ij} - \mathbf{c}_{ij})(\mathbf{p}_{ij} - \mathbf{c}_{ij})^T$$

(4.5)

As a result, the objective function in Eq. 4.4 can be rewritten as Eq. 4.6 by introducing a constraint $\mathbf{W}^T \mathbf{V}_t \mathbf{W} = \mathbf{I}$ on $\mathbf{W}$ to exclude the trivial solution and make $\mathbf{W}$ non-singular.

$$\min_{\mathbf{W}} \mathbf{Tr}(\mathbf{W}^T (\mathbf{V}_t - \mathbf{V}_f)\mathbf{W})$$

$$s.t. \mathbf{W}^T \mathbf{V}_t \mathbf{W} = \mathbf{I}$$

(4.6)

The solution $\mathbf{W}$ for optimizing the objective function in Eq. 4.6 consists of the eigenvectors of matrix $\mathbf{V}_t^{-1}(\mathbf{V}_t - \mathbf{V}_f)$. Finally, the InCS is derived from the original RGB color space by applying the learned transformation matrix $\mathbf{W}$.

It can be discovered from the objective function in Eq. 4.6 that the proposed InCS seeks to balance the criterion of minimizing the color variations among the kinship image pairs and the criterion of maximizing the color variations among the non-kinship image pairs. Therefore, the InCS is capable of capturing the color variations between the kinship images. The complete algorithm of learning the InCS is summarized as follows:

- Compute $\mathbf{m}_{p_i}$ and $\mathbf{m}_{c_i}$ by using Eq. 4.1.
- Compute $\mathbf{V}_t$ and $\mathbf{V}_f$ by using Eq. 4.5.
- Derive $\mathbf{W}$ by optimizing Eq. 4.6.
- $[C_1, C_2, C_3] = [R, G, B]\mathbf{W}$.

4.4   Properties of the InCS

This section presents two properties of the proposed InCS, namely the decorrelation property and the robustness to illumination variations. Specifically, the decorrelation property is first revealed through theoretical analyses and its connection to the verification performance is further discussed. Second, the robustness to illumination variations at both pixel level and feature level are analyzed based on the diagonal illumination model [21, 34].

4.4.1 The Decorrelation Property

For kinship verification, an important variable is the difference matrix $\mathbf{d}_i$ for the corresponding pixels between two images, which is computed as $\mathbf{d}_i = \mathbf{x}_i - \mathbf{y}_i$. The reason is that the similarity between two images can be represented in terms of $\mathbf{d}_i$ if some similarity measures, such as the CSM or the Euclidean distance, are applied. As a result, the decorrelation of the components of $\mathbf{d}_i$ can reduce the information redundancy for measuring the similarity and enhance the kinship verification performance. The following analyses demonstrate that the proposed InCS is able to decorrelates the components of $\mathbf{d}_i$, which is defined as the decorrelation property.

Given the derived transformation matrix $\mathbf{W} = [\mathbf{w}_1, \mathbf{w}_2, \mathbf{w}_3]$ where $\mathbf{w}_i \in \mathbb{R}^{3 \times 1} (i = 1, 2, 3)$, the pixel color values in vector $\mathbf{x}_{ij} = [x_{ij1}, x_{ij2}, x_{ij3}]^T$ and $\mathbf{y}_{ij} = [y_{ij1}, y_{ij2}, y_{ij3}]^T$ can be represented as $x_{iju} = \mathbf{w}_u^T \mathbf{p}_{ij}$, $y_{iju} = \mathbf{w}_u^T \mathbf{c}_{ij}$, $x_{ijv} = \mathbf{w}_v^T \mathbf{p}_{ij}$ and $y_{ijv} = \mathbf{w}_v^T \mathbf{c}_{ij}$ where $u, v = 1, 2, 3$ and $u \neq v$. Note that x_{iju} and x_{ijv} are the pixel values of the u-th component and the v-th component of $\mathbf{x}_{ij}$ in the InCS respectively. y_{iju} and y_{ijv} are defined in a similar fashion. The decorrelation property of the proposed InCS states that the u-th component (row) $\mathbf{d}_i(u)$ and the v-th component $\mathbf{d}_i(v)$ ($u, v = 1, 2, 3$ and $u \neq v$) of the difference matrix $\mathbf{d}_i$ are decorrelated.

Property 4.4.1 (The Decorrelation Property) *If each component of the color difference matrix $\mathbf{d}_i$ is centered, we have the following statistical correlation $S(u, v)$ between the u-th component (row) $\mathbf{d}_i(u)$ and the v-th component $\mathbf{d}_i(v)$ (u, v = 1, 2, 3 and u $\neq$ v) of the difference matrix $\mathbf{d}_i$.*

$$S(u, v) = \mathcal{E}\left\{(\mathbf{d}_i(u) - \mathcal{E}\{\mathbf{d}_i(u)\})(\mathbf{d}_i(v) - \mathcal{E}\{\mathbf{d}_i(v)\})^T\right\}$$
$$= 0 \tag{4.7}$$

Proof First, $\mathcal{E}\{\mathbf{d}_i(u)\}$ and $\mathcal{E}\{\mathbf{d}_i(v)\}$ are zero since each component of the color difference matrix $\mathbf{d}_i$ is centered. Second, the solution of optimizing Eq. 4.6 shows that $\mathbf{W}$ consists of the eigenvectors of matrix $\mathbf{V}_t^{-1}(\mathbf{V}_t - \mathbf{V}_f)$, which can be further proved to be the same as the eigenvectors of matrix $(\mathbf{V}_t + \mathbf{V}_f)^{-1}\mathbf{V}_f$. In other words, we have the following results

$$(\mathbf{V}_t + \mathbf{V}_f)^{-1}\mathbf{V}_f\mathbf{W} = \mathbf{W}\Lambda$$
$$\Rightarrow \mathbf{W}^T(\mathbf{V}_t + \mathbf{V}_f)\mathbf{W}\Lambda = \mathbf{W}^T\mathbf{V}_t\mathbf{W} \tag{4.8}$$
$$\Rightarrow \mathbf{W}^T(\mathbf{V}_t + \mathbf{V}_f)\mathbf{W} = \Lambda^{-1}$$

where Λ is a diagonal matrix that is composed of the eigenvalues of $(\mathbf{V}_t + \mathbf{V}_f)^{-1}\mathbf{V}_f$.

Then we have the following statistical correlation $S(u, v)$ between the u-th and v-th color components ($u, v = 1, 2, 3$ and $u \neq v$) for $\mathbf{x}_i$ and $\mathbf{y}_i$ as Eq. 4.9:

$$S(u, v) = \mathcal{E}(\mathbf{d}_i(u)\mathbf{d}_i(v)^T)$$

$$= \frac{1}{m} \sum_{i=1}^{m} \mathbf{d}_i(u)\mathbf{d}_i(v)^T$$

$$= \frac{1}{m} \sum_{i=1}^{m} (\mathbf{x}_i(u) - \mathbf{y}_i(u))(\mathbf{x}_i(v) - \mathbf{y}_i(v))^T \qquad (4.9)$$

$$= \frac{1}{m} \sum_{i=1}^{m} \sum_{j=1}^{n} (x_{iju} - y_{iju})(x_{ijv} - y_{ijv})^T$$

$$= \frac{1}{m} \mathbf{w}_u^T (V_t + V_f)\mathbf{w}_v$$

$$= 0$$

To further show the benefits of the decorrelation property, the average statistical correlation (ASC), which behaves as an indicator of the degree of decorrelation for a color space, is defined as follows.

$$ASC = \frac{1}{3} \sum_{u \neq v} |S(u, v)| \qquad (4.10)$$

where $|\cdot|$ is the absolute value. Experimental results in Sect. 4.7.2 show that in general, the smaller the ASC is, the better the performance the corresponding color space achieves.

4.4.2 Robustness to Illumination Variations

Another important property of the proposed InCS is the robustness to illumination variations. The illumination variations of an image can be modeled by the diagonal model [21, 34], which corresponds to the Lambertian reflectance model under the assumption of narrow band filters. Specifically, The diagonal model is defined as a diagonal transformation $\mathbf{L} \in \mathbb{R}^{3 \times 3}$ on the RGB values of each pixel and a shift $\mathbf{s} \in \mathbb{R}^{3 \times 1}$ as follows:

$$\begin{pmatrix} a & 0 & 0 \\ 0 & b & 0 \\ 0 & 0 & c \end{pmatrix} \begin{pmatrix} R \\ G \\ B \end{pmatrix} + \begin{pmatrix} s_1 \\ s_2 \\ s_3 \end{pmatrix} \qquad (4.11)$$

where a, b and c are the diagonal elements of matrix $\mathbf{L}$, and s_1, s_2 and s_3 are the elements of vector $\mathbf{s}$.

Based on this diagonal model, five types of common illumination variations, namely light intensity change, light intensity shift, light intensity change and shift, light color change as well as light color change and shift, can be identified [21]. First, the light intensity change assumes $a = b = c > 0$ and $s_1 = s_2 = s_3 = 0$, which means the pixel values change by a constant factor in all color components. The light intensity change is often caused by the differences of the intensity of light, shadows and shading. Second, the light intensity shift assumes $a = b = c = 1$ and $s_1 = s_2 = s_3 > 0$. The light intensity shift is mainly due to the diffuse lighting. Third, the light intensity change and shift combines the above two changes and assumes $a = b = c > 0$ and $s_1 = s_2 = s_3 > 0$. Fourth, the light color change assumes $a \neq b \neq c > 0$ and $s_1 = s_2 = s_3 = 0$, which means each color component scales independently. Finally, the light color change and shift assumes $a \neq b \neq c > 0$ and $s_1 \neq s_2 \neq s_3 > 0$.

Now we present theoretical analysis to show the following conclusions under the illumination condition that the diagonal transformation $\mathbf{L}$ applied is the same within each image pair (parent and child image in this image pair have the same $\mathbf{L}$) but different across image pairs and the shift $\mathbf{s}$ is applied similarly.

- Our InCS is robust to light intensity change and light color change.
- Our InCS becomes robust to light intensity shift, light intensity change and shift as well as light color change and shift with the help of the proposed CSM or the Euclidean distance to cancel out the shift.

Let $\mathbf{V}_t^u$ and $\mathbf{V}_f^u$ be the new true verification matrix and the new false verification matrix, respectively, after the illumination variations happen to each original image pair (I_{p_i}, I_{c_i}) by the diagonal transformation $\mathbf{L}_i$ and the shift $\mathbf{s}_i (i = 1, 2, \ldots, m)$. Note that $\mathbf{L}_i$ may be different across image pairs but are the same within each image pair, and so is $\mathbf{s}_i$. Then we have $\mathbf{V}_t^u = \mathbf{L}_i^t \mathbf{V}_t \mathbf{L}_i$ and $\mathbf{V}_f^u = \mathbf{L}_i^t \mathbf{V}_f \mathbf{L}_i$. Then the new transformation $\mathbf{W}^u$ has the relation to the original transformation $\mathbf{W}$ as $\mathbf{W} = \mathbf{L}_i \mathbf{W}^u$. As a result, the derived image pixel vectors for InCS with the illumination variations are as follows:

$$
\begin{aligned}
(\mathbf{W}^u)^T (\mathbf{L}_i^t \mathbf{p}_{ij} + \mathbf{s}_i) &= \mathbf{W}^T \mathbf{p}_{ij} + (\mathbf{W}^u)^T \mathbf{s}_i \\
(\mathbf{W}^u)^T (\mathbf{L}_i^t \mathbf{c}_{ij} + \mathbf{s}_i) &= \mathbf{W}^T \mathbf{c}_{ij} + (\mathbf{W}^u)^T \mathbf{s}_i
\end{aligned}
\tag{4.12}
$$

First, if the shift $\mathbf{s}_i$ is not considered ($s_1 = s_2 = s_3 = 0$), then the image pixel vectors for InCS with the illumination variations are the same as those without illumination variations, which means our InCS is robust to light intensity change and light color change regardless of the value of $\mathbf{L}_i$.

Second, if the shift $\mathbf{s}_i$ is considered ($s_1 \neq s_2 \neq s_3 > 0$), when our CSM or the Euclidean distance is applied, it is easy to see that the effect of the offset term $(\mathbf{W}^u)^T \mathbf{s}_i$ is canceled out when comparing the parent and child image. Then our InCS is also robust to light intensity shift, light intensity change and shift as well as light color change and shift with the help of the proposed CSM.

However, the pixel level inheritable color space cannot guarantee the robustness to illumination variations under a more general illumination condition: both the diagonal transformation $\mathbf{L}_i$ and shift $\mathbf{s}_i$ are different for parent and child image within the kinship image pair respectively. In this case, we will appeal to the feature level inheritable color space, namely the generalized InCS (GInCS) described in Sect. 4.5. We will show our GInCS is also robust to illumination variations in Sect. 4.5 under a more general illumination condition.

For empirical evaluation of the robustness of the InCS, please refer to Sect. 4.7.3, where we evaluate the kinship verification performance under different illumination conditions based on the five types of illumination variations described above.

4.5 The Generalized InCS (GInCS) Framework

Our novel InCS can be further generalized from the pixel level to the feature level to enhance generalization performance and improve robustness to image variabilities, such as illumination variations, under a more general illumination condition. Usually, shape and texture features are computed for each color component first. One example is to apply the Fisher vector method [23] to the red, green, and blue component images prior to the computation of the component images in the InCS. Figure 4.2

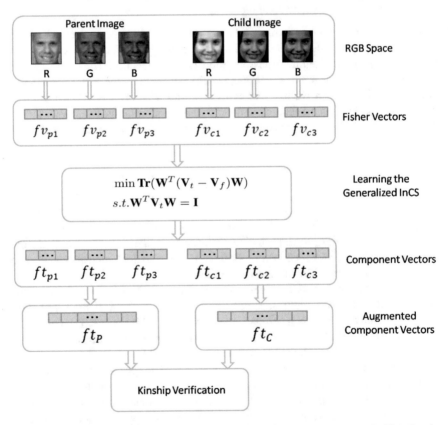

Fig. 4.2 The schematic of the generalized InCS with the Fisher vector as an example. Note that the Fisher vectors ($\mathbf{fv}_{p1}$, $\mathbf{fv}_{p2}$, $\mathbf{fv}_{p3}$, $\mathbf{fv}_{c1}$, $\mathbf{fv}_{c2}$, $\mathbf{fv}_{c3}$) are first computed from the *red*, *green*, and *blue* component images instead of the pixel vector in InCS. Then the generalized InCS is learned based on these Fisher vectors and new component vectors ($\mathbf{ft}_{p1}$, $\mathbf{ft}_{p2}$, $\mathbf{ft}_{p3}$, $\mathbf{ft}_{c1}$, $\mathbf{ft}_{c2}$, $\mathbf{ft}_{c3}$) are derived for kinship verification

shows the schematic of a generalized InCS that applies the Fisher vector first and derives the new component vectors, which define the generalized InCS. Note that other shape and texture feature extraction methods, such as Gabor wavelets [35], LBP [36], Feature LBP [37] may be applied as well to define new generalized InCS for kinship verification.

We first briefly review the Fisher vector method [23], which has been widely applied for visual recognition problems such as face recognition [38] and object recognition [23]. Specifically, let $\mathbf{X} = \{\mathbf{d}_t, t = 1, 2, \ldots, T\}$ be the set of T local descriptors (e.g. SIFT descriptors) extracted from the image. Let μ_λ be the probability density function of $\mathbf{X}$ with a set of parameters λ, then the Fisher kernel [23] is defined as $K(\mathbf{X}, \mathbf{Y}) = (\mathbf{G}_\lambda^\mathbf{X})^T \mathbf{F}_\lambda^{-1} \mathbf{G}_\lambda^\mathbf{Y}$, where $\mathbf{G}_\lambda^\mathbf{X} = \frac{1}{T} \nabla_\lambda \log[\mu_\lambda(\mathbf{X})]$, which is the gradient vector of the log-likelihood that describes the contribution of the parameters to the generation process. And $\mathbf{F}_\lambda$ is the Fisher information matrix of μ_λ. Essentially, the Fisher vector is derived from the explicit decomposition of the Fisher kernel according to the fact that the symmetric and positive definite Fisher information matrix $\mathbf{F}_\lambda$ has a Cholesky decomposition as $\mathbf{F}_\lambda^{-1} = \mathbf{L}_\lambda^T \mathbf{L}_\lambda$. Therefore, the Fisher kernel $K(\mathbf{X}, \mathbf{Y})$ can be written as a dot product between two vectors $\mathbf{L}_\lambda \mathbf{G}_\lambda^\mathbf{X}$ and $\mathbf{L}_\lambda \mathbf{G}_\lambda^\mathbf{Y}$ which are defined as the **Fisher vectors** of $\mathbf{X}$ and $\mathbf{Y}$ respectively. A diagonal closed-form approximation of $\mathbf{F}_\lambda$ [23] is often used where $\mathbf{L}_\lambda$ is just a whitening transformation for each dimension of $\mathbf{G}_\lambda^\mathbf{X}$ and $\mathbf{G}_\lambda^\mathbf{Y}$. The dimensionality of Fisher vector depends only on the number of parameters in the parameter set λ.

We now present the generalized InCS framework by using the Fisher vector as an example. First, the Fisher vector is computed for each color components in the RGB color space and denoted as $\mathbf{fv}_{p1}$, $\mathbf{fv}_{p2}$, $\mathbf{fv}_{p3}$ for parent image and $\mathbf{fv}_{c1}$, $\mathbf{fv}_{c2}$, $\mathbf{fv}_{c3}$ for child image. Second, these Fisher vectors may be taken as the input of the learning process by optimizing the objective function in Eq. 4.6 to derive the Fisher vector based generalized InCS. Then the new component vectors in generalized InCS can be derived as $\mathbf{ft}_{p1}$, $\mathbf{ft}_{p2}$, $\mathbf{ft}_{p3}$ for parent image and $\mathbf{ft}_{c1}$, $\mathbf{ft}_{c2}$, $\mathbf{ft}_{c3}$ for child image. Finally, all the color component vectors are normalized and concatenated as one augmented component vector for kinship verification.

It is easy to see that our GInCS preserves the decorrelation property since the learning process remains the same. Besides, our GInCS is also robust to illumination variations under a more general illumination condition: the diagonal transformation $\mathbf{L}$ and the shift $\mathbf{s}$ are different within and between the image pairs, respectively. First, our implementation is based on the Fisher vector, which depends on the SIFT descriptors extracted from each color component image. The SIFT descriptor is intensity shift-invariant [21] since it is based on the gradient of image which takes the derivative to cancel out the intensity shift. Therefore, our GInCS is robust to intensity shift operations of the diagonal model. Second, the SIFT descriptor is often normalized so that it is robust to light intensity change under any diagonal transformation. Third, the SIFT descriptor is computed and normalized for each color component image independently so that it is also robust to light color change [21]. As a result, our Fisher vector based generalized InCS is robust to light intensity change, light color change, light intensity shift, light intensity change and shift as well as light color change and shift under a more general illumination condition. The generalized InCS framework with Fisher vector can significantly improve the results of the pixel level InCS (see Sect. 4.7.5) and outperform other popular methods (see Sect. 4.6.1).

4.6 Experiments

Our proposed method is assessed on four representative kinship verification datasets: the KinFaceW-I dataset [3], the KinFaceW-II dataset [3], the UB KinFace dataset [2], and the Cornell KinFace dataset [1]. Example images from these datasets are shown in Fig. 4.3.

The pattern vector we implement for InCS is the concatenation of column pixels of the RGB color space. The pattern vector for the generalized InCS is the Fisher vector. In particular, first, the dense SIFT features are derived with a step size of 1 and five scale patch sizes of 2, 4, 6, 8, 10, respectively, from each color component images. The dimensionality 128 of the SIFT feature is further reduced to 64 by PCA. Then the spatial information [38] is added to the SIFT feature with 2 more dimensions and the final dimensionality of the SIFT feature is 66. Second, a Gaussian mixture model with 256 Gaussian components is computed. As a result, the Fisher vector is derived as 33792 ($2 * 256 * 66$) dimension vector. Power transformation [23] is then applied to the Fisher vector. The value of the parameter λ in the objective function is set to 0.5. Third, the Euclidean distance and our proposed color similarity measure may be

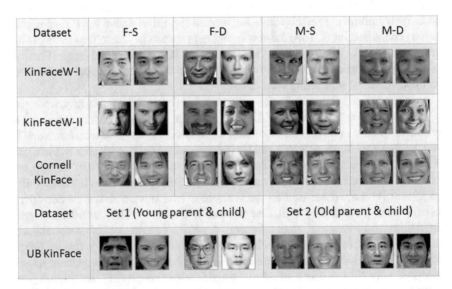

Fig. 4.3 Example images from the KinFaceW-I, the KinFaceW-II dataset, the Cornell KinFace dataset and the UB KinFace dataset. Note that the images from the Cornell KinFace dataset and the UB KinFace dataset are cropped to the same size as the KinFaceW-I and the KinFaceW-II dataset

further applied to compute the similarity between two images. Finally, a two class linear support vector machine is used to determine the kinship relations between images.

4.6.1 Experimental Results Using the KinFaceW-I and the KinFaceW-II Datasets

In KinFaceW-I dataset, each image pair was acquired from different photos whereas in KinfaceW-II, each image pair was obtained from the same photo. These datasets contain images for four kinship relations: father-son (F-S), father-daughter (F-D), mother-son (M-S), and mother-daughter (M-D). In the KinFaceW-I dataset, there are 156, 134, 116, and 127 image pairs for each of the relations defined above. In the KinFaceW-II dataset, there are 250 pairs of the images for each relation. In our experiments, we conduct 5-fold cross validation where both the KinFaceW-I dataset and the KinFaceW-II dataset are divided into five folds having the same number of image pairs. The SIFT flow [39] is extracted to pre-process the images so that the corresponding parts between two images are enhanced. Our novel color similarity measure is applied to compute the distance between two images. Experimental results on Tables 4.1 and 4.2, show that our method is able to outperform other popular methods on both datasets. In particular, our method improves upon other methods, such as GGA (generative gated autoencoders) [4], DMML [24] and MPDFL [5] by almost 6% on the KinFaceW-I dataset and almost 3% on the KinFaceW-II dataset. Another interesting finding is that the combination of the C_1 and C_3 color components (the first and the third component) of the GInCS achieves the best performance on

Table 4.1 Comparison between the GInCS and other popular methods on the KinFaceW-I dataset

Methods	F-S	F-D	M-S	M-D	Mean
CSML [31]	61.10	58.10	60.90	70.00	62.50
NCA [28]	62.10	57.10	61.90	69.00	62.30
LMNN [29]	63.10	58.10	62.90	70.00	63.30
NRML [3]	64.10	59.10	63.90	71.00	64.30
MNRML [3]	72.50	66.50	66.20	72.00	69.90
ITML [30]	75.30	64.30	69.30	76.00	71.20
DMML [24]	74.50	69.50	69.50	75.50	72.25
MPDFL [5]	73.50	67.50	66.10	73.10	70.10
GGA [4]	70.50	70.00	67.20	74.30	70.50
Anthropology [4]	72.50	71.50	70.80	75.60	72.60
GInCS	**77.25**	**76.90**	**75.82**	**81.44**	**77.85**
GInCS $(C_1 C_3)$	**77.88**	**76.88**	**77.10**	**80.64**	**78.13**

Table 4.2 Comparison between the GInCS and other popular methods on the KinFaceW-II dataset

Methods	F-S	F-D	M-S	M-D	Mean
CSML [31]	71.80	68.10	73.80	74.00	71.90
NCA [28]	73.80	70.10	74.80	75.00	73.50
LMNN [29]	74.80	71.10	75.80	76.00	74.50
NRML [3]	76.80	73.10	76.80	77.00	75.70
MNRML [3]	76.90	74.30	77.40	77.60	76.50
ITML [30]	69.10	67.00	65.60	68.30	67.50
DMML [24]	78.50	76.50	78.50	79.50	78.25
MPDFL [5]	77.30	74.70	77.80	78.00	77.00
GGA [4]	81.80	74.30	80.50	80.80	79.40
GInCS	**85.40**	**77.00**	**81.60**	**81.60**	**81.40**
GInCS (C_1)	**86.40**	**79.00**	**81.60**	**82.20**	**82.30**

the KinFaceW-I dataset. Similarly, the C_1 color component of the GInCS achieves the best results on the KinFaceW-II dataset. Note that although better results are reported in the FG 2015 Kinship Verification in the Wild challenge [6], their methods use multiple features while our method uses only one single feature and still achieves comparable results to the state-of-the-art methods.

4.6.2 Experimental Results Using the UB KinFace Dataset

The UB KinFace dataset consists of 600 images of 400 persons from 200 families. For each family, there are three images, which correspond to child, young parent and old parent. In our experiments, images are aligned according to the eye position and cropped to 64×64 such that the background information is removed and only the facial region is used for kinship verification. Two subsets of images are constructed, where set 1 consists of 200 child and young parent pairs and set 2 consists of 200 child and old parent pairs. The 5-fold cross validation is also conducted. Experimental results on Table 4.3 show that our method is able to outperform other popular methods. Particularly, our method achieves the highest kinship verification rates of 75.80 and 72.20% for both set 1 and set 2, compared with other popular methods, such as MNRML [3] and DMML [24].

4.6.3 Experimental Results Using the Cornell KinFace Dataset

The Cornell KinFace dataset contains 143 pairs of kinship images where 40, 22, 13 and 26% are the father-son (F-S), father-daughter (F-D), mother-son (M-S), and

Table 4.3 Comparison between the GInCS and other popular methods on the UB KinFace dataset

Methods	Set 1	Set 2
MCCA [40]	65.50	64.00
MMFA [40]	65.00	64.00
LDDM [41]	66.50	66.00
DMMA [42]	65.50	63.50
MNRML [3]	66.50	65.50
DMML [24]	74.50	70.00
GInCS	**75.80**	**72.20**

Table 4.4 Comparison between the GInCS and other popular methods on the Cornell KinFace dataset

Methods	F-S	F-D	M-S	M-D	Mean
MCCA [40]	71.50	65.80	73.50	63.50	68.57
MMFA [40]	71.50	66.40	73.50	64.50	68.97
LDDM [41]	73.00	66.90	74.50	67.50	70.47
DMMA [42]	71.00	65.50	73.00	65.50	68.75
MNRML [3]	74.50	68.80	77.20	65.80	71.57
DMML [24]	76.00	70.50	77.50	71.00	73.75
GInCS	**78.20**	**73.00**	**78.80**	**73.50**	**75.87**

mother-daughter (M-D) relations. In our experiments, the images are preprocessed in the same fashion as the UB KinFace dataset and 5-fold cross validation is conducted for each relation respectively. Experimental results on Table 4.4 show that our method is able to outperform other popular methods. In particular, our method achieves the highest kinship verification rates of 78.20, 73.00, 78.80, and 73.50% for all the kinship relations, respectively, when compared against the popular methods, such as MNRML [3] and DMML [24]. For the mean performance, our method also achieves the highest kinship verification rate of 75.87% among all the methods compared.

4.7 Comprehensive Analysis

This section presents a comprehensive analysis of our proposed InCS method regarding (i) the comparison with other color spaces, (ii) analysis of the decorrelation property, (iii) robustness to illumination variations, (iv) performance of the different color components of the InCS, (v) comparison between the InCS and the generalized InCS, and (vi) the generalization ability of the InCS method. All the experiments are conducted on the KinFaceW-I dataset and the KinFaceW-II datasets. The common process of the experiments is as follows. First, the concatenation of column pixels is applied for each color component image in the InCS to derive the pattern vector of each color component. Second, the pattern vectors in each component are normalized and concatenated as an augmented pattern vector. Third, the Euclidean distance or our proposed color similarity measure (CSM) is applied to calculate the similarity between different images. And finally, a linear support vector machine is utilized for kinship verification.

4.7.1 Comparative Evaluation of the InCS and Other Color Spaces

This section presents the performance of different color spaces including the RGB, rgb (normalized RGB), LSLM, I1I2I3 and CIE-XYZ color spaces. Example images of different color spaces are shown in Fig. 4.4. In order to conduct a fair comparison,

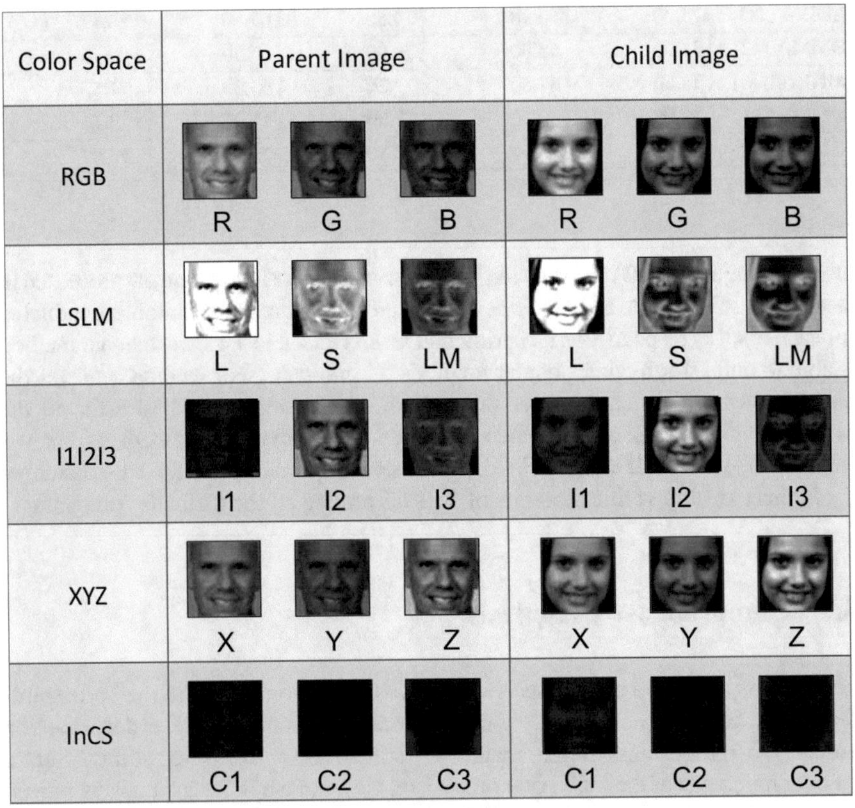

Fig. 4.4 Example images of different color spaces including RGB, LSLM, $I_1 I_2 I_3$, CIE-XYZ and InCS

Table 4.5 Comparison between the InCS and other color spaces on the KinFaceW-I dataset

Color spaces	F-S	F-D	M-S	M-D	Mean
No color	58.01	51.88	49.93	52.83	53.16
RGB	66.60	56.74	55.16	60.25	59.69
rgb	59.96	57.81	62.90	65.70	61.59
LSLM	66.31	61.20	57.37	58.28	60.79
I1I2I3	66.62	58.60	56.05	61.85	60.78
CIE-XYZ	65.32	55.60	54.73	61.05	59.18
InCS	66.35	60.06	60.76	**69.67**	**64.21**

Table 4.6 Comparison between the InCS and other color spaces on the KinFaceW-II dataset

Color spaces	F-S	F-D	M-S	M-D	Mean
No color	56.60	54.60	55.20	54.80	55.30
RGB	68.20	60.80	59.60	57.80	61.60
rgb	69.20	57.20	61.40	61.60	62.35
LSLM	70.60	61.20	63.20	60.60	63.90
I1I2I3	69.60	61.20	61.00	58.00	62.45
CIE-XYZ	67.00	60.00	60.20	58.20	61.35
InCS	**73.60**	**62.40**	**71.40**	**71.80**	**69.80**

the pattern vector is extracted by concatenating the column pixels in each color components of each image for different color spaces and the Euclidean distance is applied. Experimental results demonstrated in Tables 4.5 and 4.6 shows that the proposed InCS is able to achieve significantly better performance than the original RGB color space by a large margin of 5 and 8% on the KinFaceW-I dataset and the KinFaceW-II dataset, respectively. Besides, our proposed InCS can improve the grayscale feature (No Color) significantly by at least 11%, which provides solid proof of the effectiveness of color information for kinship verification. It is also worth noting that the room for improvement on the KinFace W-II dataset is higher due to the availability of more training samples in the KinFace W-II dataset.

4.7.2 The Decorrelation Property of the InCS Method

This section investigates the decorrelation property of our proposed InCS in terms of the average statistical correlation (ASC) defined in Eq. 4.10 for all the image pairs (including the training and testing pairs). It can be concluded from the empirical results in Tables 4.7 and 4.8 that

Table 4.7 The values of average statistical correlation of different color spaces on the KinFaceW-I dataset

Color spaces	ASC	Performance
CIE-XYZ	0.97	59.18
RGB	0.93	59.69
LSLM	0.92	60.79
I1I2I3	0.83	60.78
rgb	0.57	61.59
InCS	**0.26**	**64.21**

Table 4.8 The values of average statistical correlation of different color spaces on the KinFaceW-II dataset

Color spaces	ASC	Performance
CIE-XYZ	0.96	61.35
RGB	0.91	61.60
LSLM	0.92	63.90
I1I2I3	0.84	62.45
rgb	0.60	62.35
InCS	**0.30**	**69.80**

- Our proposed InCS is able to achieve the lowest value of ASC since it can capture the color variations between two images and decorrelates the color components of the difference matrices between two images.
- The final performances of different color spaces are quite related to the value of ASC. In general, lower value of ASC will result in better performance.

4.7.3 The Robustness of the InCS and the GInCS to Illumination Variations

To show the robustness to illumination variation for our proposed method, two groups of experiments are conducted for InCS and GInCS respectively.

In particular, the first group of experiments evaluates the robustness of InCS to illumination variations under a restricted illumination condition: for each pair of images, the same diagonal transformation L and the same shift s are applied, while for different pairs, the diagonal transformation L and the shift s may be different. We conduct the following experiments for this group: (i) First, the diagonal transformation L, whose elements fall into $(0, 1]$, is randomly generated and the elements of s are set to zero for evaluating the robustness to light intensity change (LI) and light color change (LC). As shown in Table 4.9, our InCS is indeed invariance to

Table 4.9 Performance on the KinFaceW-I and the KinFaceW-II datasets when illumination changes. "LI" means light intensity change and "LC" means light color change

	F-S	F-D	M-S	M-D	Mean
KinFaceW-I					
InCS	66.35	60.06	60.76	69.67	64.21
InCS + LI	66.35	60.06	60.76	69.67	64.21
InCS + LC	66.35	60.06	60.76	69.67	64.21
KinFaceW-II					
InCS	73.60	62.40	71.40	71.80	69.80
InCS + LI	73.60	62.40	71.40	71.80	69.80
InCS + LC	73.60	62.40	71.40	71.80	69.80

Table 4.10 Performance on the KinFaceW-I and the KinFaceW-II datasets when illumination changes. "LS" means light intensity shift, "LIs" means light intensity change and shift, and "LCs" means light color change and shift

	F-S	F-D	M-S	M-D	Mean
KinFaceW-I					
InCS	66.35	60.06	60.76	69.67	64.21
InCS + LS	66.35	60.06	60.76	69.67	64.21
InCS + LIs	66.35	60.06	60.76	69.67	64.21
InCS + LCs	66.35	60.06	60.76	69.67	64.21
KinFaceW-II					
InCS	73.60	62.40	71.40	71.80	69.80
InCS + LS	73.60	62.40	71.40	71.80	69.80
InCS + LIs	73.60	62.40	71.40	71.80	69.80
InCS + LCs	73.60	62.40	71.40	71.80	69.80

light intensity change and light color change under the restricted illumination condition. (ii) Second, the diagonal transformation $\mathbf{L}$ is randomly generated and so is the shift $\mathbf{s}$ for evaluating the robustness to light intensity shift (LS), light intensity change and shift (LIs) as well as light color change and shift (LCs). The Euclidean distance is applied. Transformed image pixel values outside the range [0, 255] are still kept but normalized later. As shown in Table 4.10, our InCS is indeed invariant to light intensity shift, light intensity change and shift as well as light color change and shift under the restricted illumination condition with the help of our CSM or the Euclidean distance. Note that all the experiments are conducted for five times and the average results are reported. The results are exactly the same for all five iterations with our CSM or the Euclidean distance. But cosine similarity produces slightly different results for each iteration.

The second group of experiments assesses the robustness of GInCS to illumination variations under a more general illumination condition: the diagonal transformation

Table 4.11 Performance of GInCS on the KinFaceW-I and the KinFaceW-II dataset for different illumination variations

	F-S	F-D	M-S	M-D	Mean
KinFaceW-I					
GInCS	77.25	76.90	75.82	81.44	77.85
GInCS + LI	77.25	76.54	77.99	80.98	78.19
GInCS + LC	76.60	76.90	76.70	80.20	77.60
GInCS + LS	77.25	76.51	77.99	81.07	78.20
GInCS + LIs	77.25	75.41	77.55	80.61	77.71
GInCS + LCs	77.25	76.15	77.12	79.81	77.58
KinFaceW-II					
GInCS	85.40	77.00	81.60	81.60	81.40
GInCS + LI	85.29	77.60	81.80	81.60	81.57
GInCS + LC	85.93	77.00	82.00	81.40	81.58
GInCS + LS	84.97	77.00	82.20	80.40	81.14
GInCS + LIs	84.97	76.80	82.40	81.20	81.34
GInCS + LCs	84.97	77.00	82.80	80.80	81.39

L and the shift **s** are different for all the images (no matter within or between the image pairs). Similarly, transformed image pixel values outside the range [0, 255] are still kept but normalized later. The results in Table 4.11 show that our GInCS method achieves illumination robustness to light intensity change, light color change, light intensity shift, light intensity change and shift as well as light color change and shift. Note that all the experiments are also conducted for five times and the average results are reported. Due to the randomness of computing the Fisher vector (random sampling of SIFT descriptors to estimate the GMM model), sometimes slightly better results are produced.

4.7.4 Performance of Different Color Components of the InCS and the GInCS

To evaluate the performance of the different color components in the InCS and the GInCS, we use C_1, C_2 and C_3 to denote these component images. Tables 4.12 and 4.13 show the experimental results of different color components and different combination of them for both InCS and GInCS. Our findings show that some components or their combinations produce better results than the combination of all the three components. In particular, as indicated by the bold highlighted results in Tables 4.12 and 4.13, the $C_1 C_3$ combination in the GInCS and the C_1 component image in the GInCS achieve the best kinship verification rate of 78.13 and 82.30%, on the KinFaceW-I and the KinFaceW-II datasets, respectively.

Table 4.12 Performance evaluation of different color components of the InCS and the GInCS on the KinFaceW-I dataset

Components	F-S	F-D	M-S	M-D	Mean
C_1 (GInCS)	78.53	76.91	74.51	81.04	77.75
C_2 (GInCS)	74.69	75.04	74.96	79.47	76.04
C_3 (GInCS)	76.29	73.89	76.68	80.21	76.77
C_1C_2 (GInCS)	76.93	76.52	76.68	81.84	77.99
C_1C_3 (GInCS)	77.88	76.88	77.10	80.64	**78.13**
C_2C_3 (GInCS)	76.29	75.76	76.25	81.04	77.34
$C_1C_2C_3$ (GInCS)	77.25	76.90	75.82	81.44	77.85
C_1 (InCS)	62.17	55.95	53.44	64.96	59.13
C_2 (InCS)	60.21	61.18	59.44	64.28	61.28
C_3 (InCS)	59.93	61.65	59.04	66.56	61.80
C_1C_2 (InCS)	65.03	59.26	58.57	67.45	62.58
C_1C_3 (InCS)	62.79	59.36	60.76	67.45	62.59
C_2C_3 (InCS)	58.68	59.36	60.82	67.05	61.48
$C_1C_2C_3$ (InCS)	66.35	60.06	60.76	69.67	64.21

Table 4.13 Performance evaluation of different color components of the InCS and the GInCS on the KinFaceW-II dataset

Components	F-S	F-D	M-S	M-D	Mean
C_1 (GInCS)	86.40	79.00	81.60	82.20	**82.30**
C_2 (GInCS)	86.40	76.40	81.80	81.00	81.40
C_3 (GInCS)	83.80	75.60	79.20	80.40	79.75
C_1C_2 (GInCS)	86.20	77.80	81.80	81.20	81.75
C_1C_3 (GInCS)	85.80	77.40	82.00	81.60	81.70
C_2C_3 (GInCS)	85.60	75.60	81.40	81.60	81.05
$C_1C_2C_3$ (GInCS)	85.40	77.00	81.60	81.60	81.40
C_1 (InCS)	63.60	54.00	68.20	60.80	61.65
C_2 (InCS)	62.40	63.20	64.80	72.00	65.60
C_3 (InCS)	69.80	62.80	72.80	71.80	69.30
C_1C_2 (InCS)	65.40	61.80	73.00	73.80	68.50
C_1C_3 (InCS)	68.40	66.20	74.40	71.60	70.15
C_2C_3 (InCS)	68.40	70.00	73.80	76.80	72.25
$C_1C_2C_3$ (InCS)	73.60	62.40	71.40	71.80	69.80

4.7.5 Comparison Between the InCS and the Generalized InCS

This section presents the comparison between the InCS and the generalized InCS. The InCS uses the pixel vector (PV)—the concatenation of column pixels while the generalized InCS applies high level features, such as the Fisher vector (FV) feature. Experimental results in Tables 4.14 and 4.15 demonstrate that the generalized InCS enhances the kinship verification performance by a large margin of more than 10%.

4.7.6 Generalization Performance of the InCS Method

This section investigates the generalization ability of the proposed InCS by conducting two experiments. First, the InCS is learned from the KinFaceW-I dataset and applied to the KinFaceW-II dataset for kinship verification. Second, the same process is conducted in the reverse order, which means the InCS is learned from the KinFaceW-II dataset and applied to the KinFaceW-I dataset. Note that the KinFaceW-I dataset and the KinFaceW-II dataset are collected in different settings. Comparing with the results reported in Tables 4.5 and 4.6, results in Table 4.16 demonstrate the generalization ability of the InCS since the performance does not vary much even though the training set is collected in a different setting.

Table 4.14 Comparison between the InCS and the GInCS on the KinFaceW-I dataset

Methods	F-S	F-D	M-S	M-D	Mean
InCS	66.35	60.06	60.76	69.67	64.21
GInCS	77.25	76.90	75.82	81.44	77.85

Table 4.15 Comparison between the InCS and the GInCS on the KinFaceW-II dataset

Features	F-S	F-D	M-S	M-D	Mean
InCS	73.60	62.40	71.40	71.80	69.80
GInCS	85.40	77.00	81.60	81.60	81.40

Table 4.16 Performance on the KinFaceW-II (KinFaceW-I) dataset when the InCS is learned from KinFaceW-I (KinFaceW-II) dataset

Training	F-S	F-D	M-S	M-D	Mean
KinFaceW-I	72.80	58.40	70.40	66.80	67.10
KinFaceW-II	63.76	63.43	58.19	67.33	63.18

4.8 Conclusion

We have presented in this chapter a novel inheritable color space (InCS) and a generalized InCS (GInCS) framework with applications to kinship verification. Specifically, we have defined a new color similarity measure (CSM), and then applied CSM to derive an innovative color model, the inheritable color space (InCS), and a generalized InCS (GInCS) framework. The InCS is learned by balancing the criterion for minimizing the distance between kinship images and the criterion for maximizing the distance between non-kinship images. Our theoretical and empirical analyses show that our proposed InCS possesses both the decorrelation property and the property of robustness to illumination variations. Furthermore, a generalized InCS framework is derived by extending the InCS from the pixel level to the feature level for improving the verification performance and the robustness to illumination variations. Extensive experiments using a number of popular kinship verification datasets show the feasibility of our proposed methods.

References

1. Fang, R., Tang, K.D., Snavely, N., Chen, T.: Towards computational models of kinship verification. In: ICIP, pp. 1577–1580 (2010)
2. Xia, S., Shao, M., Luo, J., Fu, Y.: Understanding kin relationships in a photo. IEEE Trans. Multimed. **14**(4), 1046–1056 (2012)
3. Lu, J., Zhou, X., Tan, Y.P., Shang, Y., Zhou, J.: Neighborhood repulsed metric learning for kinship verification. IEEE Trans. Pattern Anal. Mach. Intell. **36**(2), 331–345 (2014)
4. Dehghan, A., Ortiz, E.G., Villegas, R., Shah, M.: Who do i look like? Determining parent-offspring resemblance via gated autoencoders. In: CVPR, pp. 1757–1764. IEEE (2014)
5. Yan, H., Lu, J., Zhou, X.: Prototype-based discriminative feature learning for kinship verification. IEEE Trans. Cybern. (2015)
6. Lu, J., Hu, J., Liong, V.E., Zhou, X., Bottino, A., Islam, I.U., Vieira, T.F., Qin, X., Tan, X., Chen, S., Keller, Y., Mahpod, S., Zheng, L., Idrissi, K., Garcia, C., Duffner, S., Baskurt, A., Castrillon-Santana, M., Lorenzo-Navarro, J.: The FG 2015 Kinship Verification in the Wild Evaluation. In: FG 2015, pp. 1–7 (2015)
7. Alvergne, A., Faurie, C., Raymond, M.: Differential facial resemblance of young children to their parents: Who do children look like more? Evol. Hum. Behav. **28**(2), 135–144 (2007)
8. Bressan, P., Dal Martello, M.: Talis pater, talis filius: Perceived resemblance and the belief in genetic relatedness. Psychol. Sci. **13**(3), 213–218 (2002)
9. Liu, C.: Learning the uncorrelated, independent, and discriminating color spaces for face recognition. IEEE Trans. Inf. Forensics Secur. **3**(2), 213–222 (2008)
10. Liu, C., Yang, J.: ICA color space for pattern recognition. IEEE Trans. Neural Netw. **20**(2), 248–257 (2009)
11. Liu, Z., Liu, C.: Fusion of color, local spatial and global frequency information for face recognition. Pattern Recognit. **43**(8), 2882–2890 (2010)
12. Yang, J., Liu, C.: Color image discriminant models and algorithms for face recognition. IEEE Trans. Neural Netw. **19**(12), 2088–2098 (2008)
13. Yang, J., Liu, C., Zhang, L.: Color space normalization: Enhancing the discriminating power of color spaces for face recognition. Pattern Recognit. **43**(4), 1454–1466 (2010)
14. Liu, C.: Extracting discriminative color features for face recognition. Pattern Recognit. Lett. **32**(14), 1796–1804 (2011)

15. Liu, C.: Effective use of color information for large scale face verification. Neurocomputing **101**, 43–51 (2013)
16. Shih, P., Liu, C.: Comparative assessment of content-based face image retrieval in different color spaces. IJPRAI **19**(7), 873–893 (2005)
17. Banerji, S., Sinha, A., Liu, C.: New image descriptors based on color, texture, shape, and wavelets for object and scene image classification. Neurocomputing **117**, 173–185 (2013)
18. Khan, F.S., Rao, M.A., van de Weijer, J., Bagdanov, A.D., Lopez, A., Felsberg, M.: Coloring action recognition in still images. Int. J. Comput. Vis. (IJCV) **105**(3), 205–221 (2013)
19. Khan, F.S., Rao, M.A., van de Weijer, J., Bagdanov, A.D., Vanrell, M., Lopez, A.: Color attributes for object detection. In: IEEE Conference on Computer Vision and Pattern Recognition (CVPR 2012) (2012)
20. Khan, R., van de Weijer, J., Shahbaz Khan, F., Muselet, D., Ducottet, C., Barat, C.: Discriminative color descriptors. In: 2013 IEEE Conference on Computer Vision and Pattern Recognition (CVPR), pp. 2866–2873 (2013)
21. van de Sande, K.E.A., Gevers, T., Snoek, C.G.M.: Evaluating color descriptors for object and scene recognition. IEEE Trans. Pattern Anal. Mach. Intell. **32**(9), 1582–1596 (2010)
22. Zhang, J., Barhomi, Y., Serre, T.: A new biologically inspired color image descriptor. In: Computer Vision ECCV 2012, pp. 312–324. Springer, Heidelberg (2012)
23. Jegou, H., Perronnin, F., Douze, M., Sanchez, J., Perez, P., Schmid, C.: Aggregating local image descriptors into compact codes. IEEE Trans. Pattern Anal. Mach. Intell. **34**(9), 1704–1716 (2012)
24. Yan, H., Lu, J., Deng, W., Zhou, X.: Discriminative multimetric learning for kinship verification. IEEE Trans. Inf. Forensics Secur. **9**(7), 1169–1178 (2014)
25. Shahbaz Khan, F., van de Weijer, J., Vanrell, M.: Top-down color attention for object recognition. In: ICCV, pp. 979–986 (2009)
26. Yang, Y., Liao, S., Lei, Z., Yi, D., Li, S.: Color models and weighted covariance estimation for person re-identification. In: 2014 22nd International Conference on Pattern Recognition (ICPR), pp. 1874–1879 (2014)
27. Xing, E.P., Jordan, M.I., Russell, S., Ng, A.Y.: Distance metric learning with application to clustering with side-information. In: NIPS, pp. 505–512 (2002)
28. Goldberger, J., Roweis, S.T., Hinton, G.E., Salakhutdinov, R.: Neighbourhood components analysis. In: NIPS (2004)
29. Weinberger, K.Q., Saul, L.K.: Distance metric learning for large margin nearest neighbor classification. J. Mach. Learn. Res. **10**, 207–244 (2009)
30. Davis, J.V., Kulis, B., Jain, P., Sra, S., Dhillon, I.S.: Information-theoretic metric learning. In: ICML, pp. 209–216 (2007)
31. Nguyen, H., Bai, L.: Cosine similarity metric learning for face verification. ACCV **6493**, 709–720 (2011)
32. Liu, C.: The bayes decision rule induced similarity measures. IEEE Trans. Pattern Anal. Mach. Intell. **29**(6), 1086–1090 (2007)
33. Liu, C.: Discriminant analysis and similarity measure. Pattern Recognit. **47**(1), 359–367 (2014)
34. Finlayson, G.D., Drew, M.S., Funt, B.V.: Spectral sharpening: Sensor transformations for improved color constancy. J. Opt. Soc. Am. A **11**(5), 1553–1563 (1994)
35. Liu, C., Wechsler, H.: Gabor feature based classification using the enhanced fisher linear discriminant model for face recognition. IEEE Trans. Image Process., 467–476 (2002)
36. Wolf, L., Hassner, T., Taigman, Y.: Descriptor based methods in the wild. In: Real-Life Images workshop at the European Conference on Computer Vision (ECCV) (2008)
37. Gu, J., Liu, C.: Feature local binary patterns with application to eye detection. In: Neurocomputing, pp. 138–152 (2013)
38. Simonyan, K., Parkhi, O.M., Vedaldi, A., Zisserman, A.: Fisher vector faces in the wild. In: BMVC (2013)
39. Liu, C., Yuen, J., Torralba, A.: SIFT flow: Dense correspondence across scenes and its applications. IEEE Trans. Pattern Anal. Mach. Intell. **33**(5), 978–994 (2011)

40. Sharma, A., Kumar, A., Daume, H., Jacobs, D.: Generalized multiview analysis: A discriminative latent space. In: 2012 IEEE Conference on Computer Vision and Pattern Recognition (CVPR), pp. 2160–2167 (2012)
41. Mu, Y., Ding, W., Tao, D.: Local discriminative distance metrics ensemble learning. Pattern Recognit. **46**(8), 2337–2349 (2013)
42. Lu, J., Tan, Y.P., Wang, G.: Discriminative multimanifold analysis for face recognition from a single training sample per person. IEEE Trans. Pattern Anal. Mach. Intell. **35**(1), 39–51 (2013)

Chapter 5
Novel Sparse Kernel Manifold Learner for Image Classification Applications

Ajit Puthenputhussery and Chengjun Liu

Abstract This chapter presents a sparse kernel manifold learner framework for different image classification applications. First, a new DAISY Fisher vector (D-FV) feature is created by computing Fisher vectors on densely sampled DAISY features. Second, a WLD-SIFT Fisher vector (WS-FV) feature is developed by fusing the Weber local descriptors (WLD) with the SIFT descriptors, and the Fisher vectors are computed on the fused WLD-SIFT features. Third, an innovative fused Fisher vector (FFV) feature is developed by integrating the most expressive features of the D-FV, the WS-FV and the SIFT-FV features. The FFV feature is then further assessed in eight different color spaces and a novel fused color Fisher vector (FCFV) feature is computed by integrating the PCA features of the eight color FFV descriptors. Finally, we propose a sparse kernel manifold learner (SKML) method for learning a discriminative sparse representation by considering the local manifold structure and the label information based on the marginal Fisher criterion. The objective of the SKML method is to minimize the intraclass scatter and maximize the interclass separability which are defined based on the sparse criterion. The effectiveness of the proposed SKML method is assessed on different image classification datasets such as the Painting-91 dataset for computational art painting classification, the CalTech-101 dataset for object categorization, and the 15 Scenes dataset for scene classification. Experimental results show that our proposed method is able to achieve better performance than other popular image descriptors and learning methods for different visual recognition applications.

5.1 Introduction

In recent years, content-based image classification applications have expanded greatly due to the advent of an image era of big data resulting in the availability of large number of color images on the Internet. An important fundamental initial

A. Puthenputhussery (✉) · C. Liu (✉)
New Jersey Institute of Technology, Newark, NJ 07102, USA
e-mail: avp38@njit.edu

C. Liu
e-mail: chengjun.liu@njit.edu

© Springer International Publishing AG 2017
C. Liu (ed.), *Recent Advances in Intelligent Image Search and Video Retrieval*,
Intelligent Systems Reference Library 121, DOI 10.1007/978-3-319-52081-0_5

step for any image classification and visual recognition based system is the computation of feature descriptors that capture different kinds of information from the image. After the creation of the feature descriptors, different learning methods such as manifold learning, sparse coding are applied to increase the discriminative ability of the classification system to achieve better recognition performance. Several manifold learning and sparse coding methods have gained increasing attention for different image classification applications such as face recognition [19, 20, 22, 23], scene recognition [9, 11, 15, 17, 38, 55], object classification [15, 38, 41, 51, 55], and action recognition [1, 10, 36, 44, 49, 54].

As the human visual system is much more efficient and robust in classifying different visual elements in an image, any image classification system based on the human visual system is likely to achieve good performance for classification tasks. Different visual aspects in an image such as color, edges, shape, intensity and orientation of objects help humans to identify and discriminate between images. Pioneer works in cognitive psychology believe that the human visual cortex represent images as sparse structures as it provides an efficient representation for later stages of visual processing [26, 27]. A sparse representation of a data-point can be represented as a linear combination of a small set of basis vectors allowing efficient storage and retrieval of data. Another advantage of sparse representation is that it adapts to varying level of information in the image since it provides a distributed representation of an image. Therefore, we introduce a hybrid feature extraction method to capture different kinds of information from the image and propose a discriminative sparse coding method based on a manifold learning algorithm to learn an efficient and robust discriminative sparse representation of the image.

In this chapter, we first present novel DAISY Fisher vector (D-FV) and Weber-SIFT Fisher vector (WS-FV) features in order to handle the inconsistencies and variations of different visual classes in images. In particular, the D-FV feature enhances the Fisher vector feature by fitting dense DAISY descriptors [35] to a parametric generative model. We then develop the WS-FV by integrating Weber local descriptors [5] with SIFT descriptors and Fisher vectors are computed on the sampled WLD-SIFT features. An innovative fused Fisher vector (FFV) is proposed by fusing the principal components of D-FV, WS-FV and SIFT-FV (S-FV) features. We then assess our FFV feature in eight different color spaces and propose several color FFV features. The descriptors that are defined in different color spaces provide stability against image variations such as rotation, viewpoint, clutter and occlusions [34] which are essential for classification of images. We further extend this concept by integrating the FFV features in eight different color spaces to form a novel fused color Fisher vector (FCFV) feature. Finally, we use a sparse kernel manifold learner (SKML) method to learn a discriminative sparse representation by integrating the discriminative marginal Fisher analysis criterion to the sparse representation criterion. In particular, new intraclass compactness and interclass separability are define based on the sparse representation criterion under the manifold learning framework. The objective of the SKML method is to increase the interclass distance between data-points belonging to different classes and decrease the intraclass distance between data-points of the same class. The SKML method can efficiently calculate a global shared dictionary

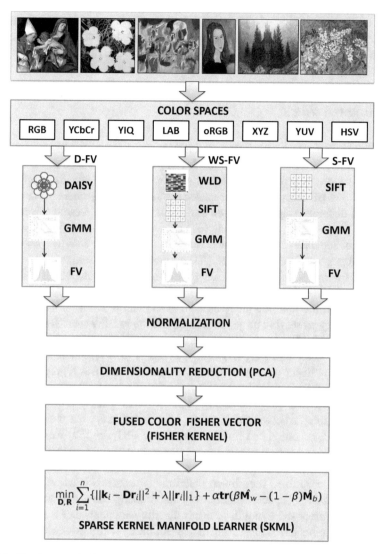

Fig. 5.1 The framework of our proposed SKML method

without the need for computation of sub-dictionaries and hence is suitable for large datasets. The framework of our proposed SKML method is illustrated in Fig. 5.1. Experimental results show that the proposed approach achieves better results compared to other popular image descriptors and state-of-the-art deep learning methods on different image classification datasets.

The rest of this chapter is organized in the following manner. In Sect. 5.2, we provide a brief discussion of some related work on image descriptors, manifold learning methods, sparse coding algorithms and screening rules. Section 5.3 describes the details of the computation of different Fisher vector features and the SKML method. We present an extensive experimental evaluation and analysis of the proposed SKML method for different classification datasets in Sects. 5.4 and 5.5 concludes the paper.

5.2 Related Work

The different cues by which humans can identify and discriminate between different visual elements in an image are the local, spatial, intensity, shape and color information, therefore they play an important role for image classification applications. The complete LBP descriptor was proposed by Guo et al. [12] where the center pixel and a local difference sign-magnitude transform is used to represent sub-regions in an image. The work of van de Sande [31] showed that fusion of SIFT descriptors computed in separate channels of different color spaces can achieve good classification performance. Shechtman et al. [32] proposed the self-similarity descriptor to capture the internal semantics of an image based on similarity matching of the layout of different visual elements in an image.

Several manifold learning methods such as the marginal Fisher analysis (MFA) [45], the locality preserving projections [13], and the locality sensitive discriminant analysis (LSDA) [3], have been widely used to preserve data locality in the embedding space. The MFA method [45] proposed by Yan et al. overcomes the limitations of the traditional linear discriminant analysis method and uses a graph embedding framework for supervised dimensionality reduction. Cai et al. [3] proposed the LSDA method that discovers the local manifold structure by finding a projection which maximizes the margin between data points from different classes at each local area. A sparse ULDA (uncorrelated linear discriminant analysis) [53] was proposed by Zhang et al. where a sparse constraint was integrated to the ULDA transformation resulting in a l_1 minimization problem with orthogonality constraint. Li et al. [18] presented the constrained multilinear discriminant analysis (CMDA) for learning the optimal tensor subspace by iteratively maximizing the scatter ratio criterion. Yan et al. [46] proposed a multitask linear discriminant analysis for effective view invariant descriptors based on sharing of self-similarity matrices. A clustering based discriminant analysis was proposed by Chen et al. [6] to overcome the shortcoming of the Fisher linear discriminant analysis and preserve the local structure information.

The different dictionary based sparse coding methods for learning a discriminative representation can be categorized into two types. The first category combines multiple sub-dictionaries of every class in order to form a unified dictionary with improved discriminatory power. The second category co-trains the sparse model and the discriminative criterion together resulting in better interclass separability between the classes achieved by adding a discriminative term to the objective function. Zhang et al. [52] proposed a method to co-train the discriminative dictionary, sparse representation as well as the linear classifier using a combined objective function that is optimized by the discriminative KSVD (D-KSVD). The D-KSVD method attempts to learn an over-complete dictionary by directly incorporating the labels in the learning stage. Jiang et al. [15] proposed a LC-KSVD method that improves upon the method introduced in [52] by introducing a label consistent regularization term. In particular, the LC-KSVD is able to jointly learn the dictionary, discriminative coding parameters and the classifier parameters. The D-KSVD and LC-KSVD methods are closely tied to linear classifiers, which discourages the use of non-linearity that can often obtain better results. Zhou et al. [56] presented a Joint Dictionary Learning (JDL) method that jointly learns both a commonly shared dictionary and class-specific sub-dictionaries to enhance the discriminative ability of the learned dictionaries. Yang et al. [47, 48] proposed a Fisher Discrimination Dictionary Learning (FDDL) method, which learns a structured dictionary that consists of a set of class-specific sub-dictionaries. The JDL and FDDL methods are time-consuming when the number of classes is large and may reduce the performance when the number of the training images for each class is small.

To increase the computational efficiency of different sparse learning models and to improve the scalability to large datasets, researchers use the screening rules. Wang et al. [39] proposed a sparse logistic regression screening rule to identify the zero components in the solution vector to effectively discard features for l_1 regularized logistic regression. Xiang et al. [42, 43] presented a dictionary screening rule to select subset of codewords to use in Lasso optimization and derived fast Lasso screening tests to find which datapoints and codewords are highly correlated. A new set of screening rules for the Lasso problem was developed by Wang et al. [37] that uses non-expansiveness of the projection operator to effectively identify inactive predictors of the Lasso problem.

5.3 Novel Sparse Kernel Manifold Learner Framework

5.3.1 Fisher Vector

We briefly review the Fisher vector which is widely applied for visual recognition problems such as face detection and recognition [33] and object recognition [14]. Fisher vector describes an image by what makes it different from other images [14] and focuses only on the image specific features. Particularly, let $\mathbf{X} = \{\mathbf{d}_t, t = 1, 2, ..., T\}$ be the set of T local descriptors extracted from the image. Let μ_λ be the probability density function of $\mathbf{X}$ with parameter λ, then the Fisher kernel [14] is defined as follows:

$$K(\mathbf{X}, \mathbf{Y}) = (\mathbf{G}_\lambda^X)^T \mathbf{F}_\lambda^{-1} \mathbf{G}_\lambda^Y \tag{5.1}$$

where $\mathbf{G}_\lambda^X = \frac{1}{T} \nabla_\lambda \log_{\mu_\lambda}(\mathbf{X})$, which is the gradient vector of the log-likelihood that describes the contribution of the parameters to the generation process. And $\mathbf{F}_\lambda$ is the Fisher information matrix of μ_λ.

Since $\mathbf{F}_\lambda^{-1}$ is positive semi-definite, it has a Cholesky decomposition as $\mathbf{F}_\lambda^{-1} = \mathbf{L}_\lambda^T \mathbf{L}_\lambda$ [14]. Therefore, the kernel $K(\mathbf{X}, \mathbf{Y})$ can be written as a dot product of the normalized vectors $\mathscr{G}_\lambda$, obtained as $\mathscr{G}_\lambda^X = \mathbf{L}_\lambda \mathbf{G}_\lambda^X$ where $\mathscr{G}_\lambda^X$ is the Fisher vector of $\mathbf{X}$ [14].

5.3.2 DAISY Fisher Vector (D-FV)

In this section, we present a new innovative DAISY Fisher vector (D-FV) feature where the Fisher vectors are computed on the densely sampled DAISY descriptors. The DAISY descriptors are suitable for dense computation and offer precise local-ization and rotational robustness [35], therefore the provide improved performance and better accuracy than some other image descriptors for classification. The DAISY descriptor [35] $\mathscr{D}(u_0, v_0)$ for location (u_0, v_0) is defined as follows:

$$
\begin{aligned}
\mathscr{D}(u_0, v_0) = [\tilde{\mathbf{h}}_{\Sigma_1}^T(u_0, v_0), \\
\tilde{\mathbf{h}}_{\Sigma_1}^T(\mathbf{I}_1(u_0, v_0, R_1)), ..., \tilde{\mathbf{h}}_{\Sigma_1}^T(\mathbf{I}_T(u_0, v_0, R_1)), ..., \\
\tilde{\mathbf{h}}_{\Sigma_Q}^T(\mathbf{I}_1(u_0, v_0, R_Q)), ..., \tilde{\mathbf{h}}_{\Sigma_Q}^T(\mathbf{I}_T(u_0, v_0, R_Q))]^T
\end{aligned}
\tag{5.2}
$$

where $\mathbf{I}_j(u, v, R)$ is the location with distance R from (u, v) in the direction given by j, Q represents the number of circular layers, and $\tilde{\mathbf{h}}_\Sigma(u, v)$ is the unit norm of vector containing Σ-convolved orientation maps in different directions. The sampled descriptors are fitted to a Gaussian Mixture Model (GMM) with 256 parameters. The Fisher vectors are then encoded as derivatives of log-likelihood of the model.

5.3.3 Weber-SIFT Fisher Vector (WS-FV)

In this section, we propose a new Weber-SIFT Fisher vector (WS-FV) feature that computes the Fisher vector on Weber local descriptor (WLD) integrated with SIFT features so as to encode the color, local, relative intensity and gradient orientation information from an image. The WLD [5] is based on the Weber's law which states that the ratio of increment threshold to the background intensity is a constant. The descriptor contains two components differential excitation [5] and orientation [5] which are defined as follows.

$$
\xi(x_c) = \arctan[\frac{v_s^{00}}{v_s^{01}}] \ and \ \theta(x_c) = \arctan(\frac{v_s^{11}}{v_s^{10}})
\tag{5.3}
$$

where $\xi(x_c)$ is the differential excitation and $\theta(x_c)$ is the orientation of the current pixel x_c, $x_i (i = 0, 1, ...p - 1)$ denotes the i-th neighbours of x_c and p is the number of neighbors, v_s^{00}, v_s^{01}, v_s^{10} and v_s^{11} are the output of filters f_{00}, f_{01}, f_{10} and f_{11} respectively. The WLD descriptor extracts the relative intensity and gradient information similar to humans perceiving the environment, therefore provides stability against noise and illumination changes. A parametric generative model is trained by fitting to the WLD-SIFT features and Fisher vectors are extracted by capturing the average first order and second order differences between the computed features and each of the GMM centers.

5.3.4 Fused Color Fisher Vector (FCFV)

In this section, we first present an innovative fused Fisher vector (FFV) feature that fuses the most expressive features of the D-FV, WS-FV and SIFT-FV features. The most expressive features are extracted by means of Principal Component Analysis (PCA) [8]. Particularly, let $\mathbf{X} \in \mathbb{R}^N$ be a feature vector with covariance matrix $\boldsymbol{\Sigma}$ given as follows: $\boldsymbol{\Sigma} = \mathbb{E}[(\mathbf{X} - \mathbb{E}(\mathbf{X}))][(\mathbf{X} - \mathbb{E}(\mathbf{X}))]^T$ where T represents transpose operation and $\mathbb{E}(.)$ represents expectation. The covariance matrix can be factorized as follows [8]: $\boldsymbol{\Sigma} = \boldsymbol{\phi}\boldsymbol{\Lambda}\boldsymbol{\phi}^T$ where $\boldsymbol{\Lambda} = diag[\lambda_1, \lambda_2, \lambda_3, ..., \lambda_N]$ is the diagonal eigenvalue matrix and $\boldsymbol{\phi} = [\boldsymbol{\phi}_1\boldsymbol{\phi}_2\boldsymbol{\phi}_3....\boldsymbol{\phi}_N]$ is the orthogonal eigenvector matrix. The most expressive features of $\mathbf{X}$ is given by a new vector $\mathbf{Z} \in \mathbb{R}^K : \mathbf{Z} = \mathbf{P}^T\mathbf{X}$ where $\mathbf{P} = [\boldsymbol{\phi}_1\boldsymbol{\phi}_2\boldsymbol{\phi}_3....\boldsymbol{\phi}_K]$ and $K < N$:

We incorporate color information to our proposed feature as the color cue provides powerful discriminating information in pattern recognition and can be very effective for face, object, scene and texture classification [21, 34]. The descriptors defined in different color spaces provide stability against illumination, clutter, viewpoint and occlusions [34]. To derive the proposed FCFV feature, we first compute the D-FV, WS-FV and SIFT-FV in the eight different color spaces namely RGB, YCbCr, YIQ, LAB, oRGB, XYZ, YUV and HSV. Figure 5.2 shows the component images of a sample image from the Painting-91 dataset in different color spaces used in this paper. For each color space, we derive the FFV by fusing the most expressive features of D-FV, WS-FV and SIFT-FV for that color space. We then reduce the dimensionality of the eight FFV features using PCA, which derives the most expressive features with respect to the minimum square error. We finally concatenate the eight FFV features and normalize to zero mean and unit standard deviation to create the novel FCFV feature.

5.3.5 Sparse Kernel Manifold Learner (SKML)

In this section, we present a sparse kernel manifold learner (SKML) to learn a compact discriminative representation by considering the local manifold structure and the label information. In particular, new within class scatter and between class scatter matrices are defined constrained by the marginal Fisher criterion [45] and the

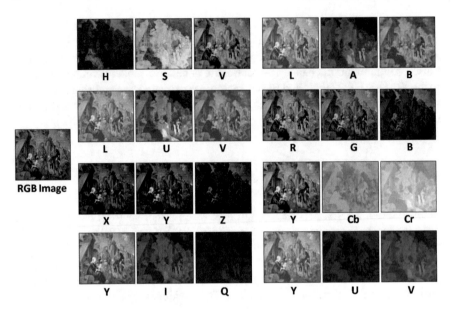

Fig. 5.2 The color component images of a sample image from the Painting-91 dataset in different colorspaces

sparse criterion so as to increase the interclass separability and the intraclass compactness based on a manifold learning framework. A discriminative term is then integrated to the representation criterion of the sparse model so as to improve the pattern recognition performance.

The features used as input for the SKML method are the FCFV features extracted from the image. Given the Fisher kernel matrix $\mathbf{K} = [\mathbf{k}_1, \mathbf{k}_2, ..., \mathbf{k}_n] \in \mathbb{R}^{m \times n}$, which contains n samples in a m dimensional space, let $\mathbf{D} = [\mathbf{d}_1, \mathbf{d}_2, ..., \mathbf{d}_b] \in \mathbb{R}^{m \times b}$ denote the dictionary that represents b basis vectors and $\mathbf{R} = [\mathbf{r}_1, \mathbf{r}_2, ..., \mathbf{r}_n] \in \mathbb{R}^{b \times n}$ denote the sparse representation matrix which represents the sparse representation for m samples. The coefficient $\mathbf{r}_i$ in the sparse representation $\mathbf{R}$ correspond to the items in the dictionary $\mathbf{D}$.

In the proposed SKML method, we jointly optimize the sparse representation criterion and the marginal Fisher analysis criterion to derive the dictionary $\mathbf{D}$ and sparse representation $\mathbf{S}$ from the training samples. The objective of the marginal Fisher analysis criterion is to minimize the intraclass compactness and maximize the interclass separability. We define new discriminative intraclass compactness $\hat{\mathbf{M}}_w$ based on the sparse criterion as follows:

$$\hat{\mathbf{M}}_w = \sum_{i=1}^{n} \sum_{(i,j) \in N_k^w(i,j)} (\mathbf{r}_i - \mathbf{r}_j)(\mathbf{r}_i - \mathbf{r}_j)^T \tag{5.4}$$

where $(i, j) \in N_k^w(i, j)$ represents the (i, j) pairs where sample $\mathbf{k}_i$ is among the nearest neighbors of sample $\mathbf{k}_j$ of the same class or vice versa.

And the discriminative interclass separability $\hat{\mathbf{M}}_b$ is defined as:

$$\hat{\mathbf{M}}_b = \sum_{i=1}^{m} \sum_{(i,j) \in N_k^b(i,j)} (\mathbf{r}_i - \mathbf{r}_j)(\mathbf{r}_i - \mathbf{r}_j)^T \tag{5.5}$$

where $(i, j) \in N_k^b(i, j)$ represents nearest (i, j) pairs among all the (i, j) pairs between samples $\mathbf{k}_i$ and $\mathbf{k}_j$ of different classes.

Therefore, we define the modified optimization criterion as:

$$\min_{\mathbf{D},\mathbf{R}} \sum_{i=1}^{n} \{||\mathbf{k}_i - \mathbf{Dr}_i||^2 + \lambda ||\mathbf{r}_i||_1\} + \alpha \, \mathbf{tr} \, (\beta \hat{\mathbf{M}}_w - (1 - \beta)\hat{\mathbf{M}}_b) \tag{5.6}$$
$$s.t. ||\mathbf{d}_j|| \leq 1, (j = 1, 2, ..., b)$$

where $\mathbf{tr(.)}$ denotes the trace of a matrix, the parameter λ controls the sparsity term, the parameter α controls the discriminatory term, the parameter β balances the contributions of the discriminative intraclass compactness $\hat{\mathbf{M}}_w$ and interclass separability $\hat{\mathbf{M}}_b$.

Let $\mathbf{L} = \mathbf{l}_1, \mathbf{l}_2, ..., \mathbf{l}_t$ are the test data matrix and t be the number of test samples, then as the dictionary $\mathbf{D}$ is already learned, the discriminative sparse representation for the test data can be derived by optimizing the following criterion:

$$\min_{S} \sum_{i=1}^{t} \{||\mathbf{l}_i - \mathbf{Ds}_i||^2\} + \lambda ||\mathbf{s}_i||_1 \tag{5.7}$$

The discriminative sparse representation for the test data is defined as $\mathbf{S} = [\mathbf{s}_1, ..., \mathbf{s}_t] \in \mathbb{R}^{b \times t}$ and has both the sparseness and discriminative information since we learn the dictionary from the the training process.

5.4   Experiments

We assess the performance of our proposed SKML method on three different image classifiction datasets namely the Painting-91 dataset [16], the CalTech 101 dataset [17] and the 15 Scenes dataset [7]. Figure 5.3 shows some sample images from different visual recognition datasets used for evaluation.

(a) Painting-91 Dataset (Fine Art Painting Categorization)

(b) Fifteen Scene Categories Dataset (Scene Recognition)

(c) Caltech 101 Dataset (Object Recognition)

Fig. 5.3 Some sample images from different visual recognition datasets used for evaluation of the proposed SKML method

5.4.1 Painting-91 Dataset

This section assesses the effectiveness of our proposed features on the challenging Painting-91 dataset [16]. The dataset contains 4266 fine art painting images by 91 artists. The images are collected from the Internet and each artist has variable number of images ranging from 31 (Frida Kahlo) to 56 (Sandro Boticelli). The dataset classifies 50 painters to 13 styles with style labels as follows: abstract expressionism (1), baroque (2), constructivism (3), cubbism (4), impressionism (5), neoclassical (6), popart (7), post-impressionism (8), realism (9), renaissance (10), romanticism (11), surrealism (12) and symbolism (13).

Art painting categorization is a challenging task as the variations in subject matter, appearance, theme and styles are large in the art paintings of the same artists. Another issue is that the similarity gap between paintings of the same styles is very small due to common influence or origin. In order to effectively classify art paintings, key aspects such as texture form, brush stroke movement, color, sharpness of edges, color balance, contrast, proportion, pattern, etc. have to be captured [30]. Painting art images are different from photographic images due to the following reasons: (i) Texture, shape and color patterns of different visual classes in art images (say, a multicolored face or a disproportionate figure) are inconsistent with regular photographic images. (ii) Some artists have a very distinctive style of using specific colors (for ex: dark shades, light

Table 5.1 Classification performance of our FFV feature in different color spaces and their fusion on the Painting-91 dataset

Feature	Artist CLs	Style CLs
RGB-FFV	59.04	66.43
YCbCr-FFV	58.41	65.82
YIQ-FFV	59.22	66.26
LAB-FFV	49.30	59.98
oRGB-FFV	57.50	65.46
XYZ-FFV	56.41	64.32
YUV-FFV	57.70	64.25
HSV-FFV	51.43	60.58
SKML	**63.09**	**71.67**

shades etc.) and brush strokes resulting in art images with diverse background and visual elements. The proposed SKML framework uses FCFV features which captures different kinds of information from the painting image and the SKML method aims to improve the discrimination between classes essential for computational fine art painting categorization.

5.4.1.1   Performance in Different Color Spaces

This section demonstrates the performance of our proposed SKML feature in eight different color spaces namely RGB, YCbCr, YIQ, LAB, oRGB, XYZ, YUV and HSV as shown in Table 5.1. Among the single color descriptors, the YIQ-FFV feature performs the best with classification accuracy of 59.22% for the artist classification task whereas the RGB-FFV feature gives the best performance of 66.43% for the style classification task. The SKML feature is computed by using a sparse representation model on the fusion of the FFV features in eight different color spaces and it achieves the best performance in both artist and style classification re-emphasizing the fact that adding color information is particularly suitable for classification of art images.

5.4.1.2   Artist and Style Classification

This section evaluates the performance of our proposed method on the task of artist and style classification. The artist classification is a task wherein a painting image has to be classified to its respective artist whereas the style classification task is to assign a style label to the painting image. The artist classification task contains 91 artists with 2275 train and 1991 test images. Similarly, the style classification task contains 13 style categories with 1250 train and 1088 test images. Table 5.2 shows the comparison of the proposed SKML feature with other state-of-the-art

Table 5.2 Comparison of the proposed feature with popular image descriptors for artist and style classification task of the Painting-91 dataset

No.	Feature	Artist CLs	Style CLs
1	LBP [16, 25]	28.5	42.2
2	Color-LBP [16]	35.0	47.0
3	PHOG [2, 16]	18.6	29.5
4	Color-PHOG [16]	22.8	33.2
5	GIST [16]	23.9	31.3
6	Color-GIST [16]	27.8	36.5
7	SIFT [16, 24]	42.6	53.2
8	CLBP [12, 16]	34.7	46.4
9	CN [16, 40]	18.1	33.3
10	SSIM [16]	23.7	37.5
11	OPPSIFT [16, 31]	39.5	52.2
12	RGBSIFT [16, 31]	40.3	47.4
13	CSIFT [16, 31]	36.4	48.6
14	CN-SIFT [16]	44.1	56.7
15	Combine (1–14) [16]	53.1	62.2
16	**SKML**	**63.09**	**71.67**

image descriptors. The color LBP descriptor [16] is calculated by fusing the LBP descriptors computed on the R,G and B channels of the image. Similar strategy is used to compute the color versions of PHOG and GIST descriptor. The opponent SIFT [31] for the painting image is computed by first converting the image to the opponent color space and then fusing the SIFT descriptors calculated for every color channel. The SSIM (self similarity) descriptor [16] is computed using a correlation map to estimate the image layout. The combination of all image descriptors listed in Table 5.2 gives a classification accuracy of 53.1% and 62.2% for the artist and style classification tasks respectively. Experimental results show that our proposed SKML feature significantly outperforms popular image descriptors and their fusion, and achieves the classification performance of 63.09% and 71.67% for artist and style classification respectively.

Table 5.3 shows the performance of the proposed SKML features compared with state-of-the-art deep learning methods. MSCNN [29] stands for multi-scale convolutional neural network which extracts features in different scales using multiple CNNs. The cross layer convolutional neural network (CNN F) [28] computes features from multiple layers of CNN to improve discriminative ability instead of only extracting from the top-most layer. The best performing CNN for the artist classification and style classification is MSCNN-1 [29] and MSCNN-2 [29] respectively. Our proposed SKML method achieves better result compared to state-of-the-art deep learning methods such as multi scale CNN and cross layer CNN.

Table 5.3 Comparison of the proposed feature with state-of-the-art deep learning methods for artist and style classification task of the Painting-91 dataset

No.	Feature	Artist CLs	Style CLs
1	MSCNN-0 [29]	55.15	67.37
2	MSCNN-1 [29]	58.11	69.69
3	MSCNN-2 [29]	57.91	70.96
4	MSCNN-3 [29]	–	67.74
5	CNN F_1 [28]	55.40	68.20
6	CNN F_2 [28]	56.25	68.29
7	CNN F_3 [28]	56.40	68.57
8	CNN F_4 [28]	56.35	69.21
9	CNN F_5 [28]	56.35	69.21
10	**SKML**	**63.09**	**71.67**

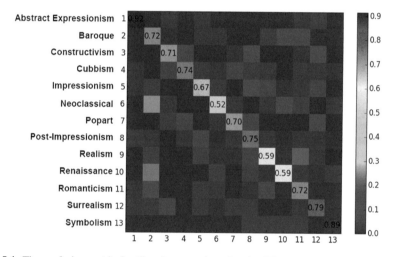

Fig. 5.4 The confusion matrix for 13 style categories using the SKML feature

Figure 5.4 shows the confusion matrix for the 13 style categories using the SKML feature where the rows denote the actual classes while the columns denote the assigned classes. It can be observed that the best classified categories are 1 (abstract expressionism) and 13 (symbolism) with classification rates of 92% and 89% respectively. The most difficult category to classify is category 6 (neoclassical) as there are large confusions between the styles baroque and neoclassical. Similarly, the other categories that create confusion are the styles baroque and renaissance.

Table 5.4 Art movement associated with different art styles

Art movement	Art style
Renaissance	Renaissance
Post renaissance	Baroque, neoclassical, romanticism, realism
Modern art	Popart, impressionism, post impressionism, surrealism, cubbism, symbolism, constructivism, abstract expressionism

5.4.1.3 Comprehensive Analysis of Results

Table 5.4 shows the art movements associated with different art styles. Interesting patterns can be observed from the confusion diagram in Fig. 5.4. The art styles within an art movement show higher confusions compared to the art styles between the art movement periods. An art movement is a specific period of time wherein an artist or group of artists follow a specific common philosophy or goal. It can be seen that there are large confusions for the styles baroque and neoclassical. Similarly, the style categories romanticism and realism have confusions with style baroque. The style categories baroque, neoclassical, romanticism and realism belong to the same art movement period—post renaissance. Similarly, popart paintings have confusions with style category surrealism within the same art movement but none of the popart paintings are misclassified as baroque or neoclassical. The only exception to the above observation is the style categories renaissance and baroque as even though they belong to different art movement period, there are large confusions between them. The renaissance and baroque art paintings have high similarity as the baroque style evolved from the renaissance style resulting in few discriminating aspects between them [30].

5.4.1.4 Artist Influence

In this section, we analyze the influence an artist can have over other artists. We find the influence among artists by looking at similar characteristics between the artist paintings. Artist influence may help us to find new connections among artists during different art movement period and also understand the influence among different art movement periods. In order to calculate the artist influence, we calculate the correlation score between the paintings of different artists. Let $\mathbf{a}_{ik}$ denote the feature vector representing the painting by artist k where $i = 1, .., n_k$ and let n_k be the total number of paintings by artist k. We calculate $\mathbf{A}_k$ which is the average of the feature vector of all paintings by artist k. We then compute a correlation matrix by comparing the average feature vector of each artist with all other artists. Finally, clusters are defined for artists with high correlation score. Figure 5.5 show the artist influence cluster graph with correlation threshold of 0.70.

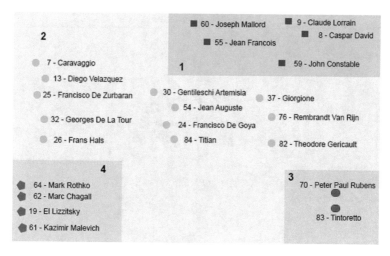

Fig. 5.5 The artist influence cluster graph for the Painting-91 dataset

Interesting observations can be deduced from Fig. 5.5. Every cluster can be associated with a particular style and time period. Cluster 1 shows artists with major contributions to the styles realism and romanticism and they belong to the post renaissance art movement period. Cluster 2 has the largest number of artists associated with the styles renaissance and baroque. Cluster 3 represents artists for the style Italian renaissance that took place in the 16th century. And cluster 4 shows artists associated with style abstract expressionism in the modern art movement period (late 18th–19th century).

We further show the k-means clustering graph with cosine distance to form clusters of similar artists. Figure 5.6 shows the artist influence graph clusters for paintings of all artists with k set as 8. First the average of the feature vector of all paintings of an artist is calculated as described above. We then apply k-means clustering algorithm with k set as 8. The artist influence graph is plotted using the first two principal components of the average feature vector. The results of Fig. 5.5 have high correlation with the results of the artist influence cluster graph in Fig. 5.6.

5.4.1.5 Style Influence

In this section, we study the style influence so as to find similarities between different art styles and understand the evolution of art styles in different art movement periods. The style influence is calculated in a similar manner as the artist influence. First, we calculate the average of the feature vector of all paintings for a style. We then apply k-means clustering method with cosine distance to form clusters of similar styles. We set the number of clusters as 3 based on the different art movement periods.

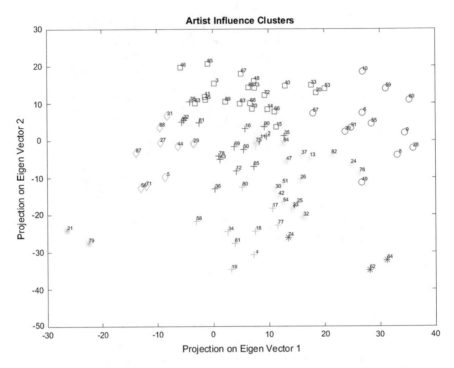

Fig. 5.6 The artist influence cluster graph using k means clustering for the Painting-91 dataset

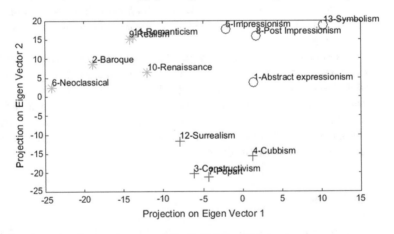

Fig. 5.7 The style influence cluster graph for the Painting-91 dataset

The style influence graph is plotted using the first two principal components of the average feature vector.

Figure 5.7 shows the style influence graph clusters with k set as 3. Cluster 1 contains the styles of the post renaissance art movement period with the only exception of style renaissance. The reason for this may be due the high similarity between styles baroque and renaissance as the style baroque evolved from the style renaissance [30]. The styles impressionism, post impressionism and symbolism in cluster 2 show that there are high similarities between these styles in the modern art movement period as the three styles have a common french and belgian origin. Similarly, styles constructivism and popart in cluster 3 show high similarity in the style influence cluster graph.

We further show the results based on the correlation matrix computed by comparing the average feature vector of all paintings of each style with all other styles. We set the correlation threshold as 0.7.

Renaissance ⟹ Baroque, Neoclassical

Romanticism ⟹ Realism

Impressionism ⟹ Post impressionism

Constructivism ⟹ Popart

The results are in good agreement with the style influence cluster graph and support the observation that the art styles within an art movement show higher similarity compared to the art styles between the art movement periods. The styles baroque and neoclassical belong to the same art movement period and the style baroque has evolved from the style renaissance. Similarly, other styles belong to the modern art movement period. It can be observed from the style influence cluster graph that the style pairs romanticism:realism, impressionism:post impressionism and constructivism:popart are plotted close to each other in the graph indicating high similarity between these styles.

5.4.2 Fifteen Scene Categories Dataset

The fifteen scene categories dataset [17] contains 4485 images from fifteen scene categories namely, office, kitchen, living room, bedroom, store, industrial, tall building, inside cite, street, highway, coast, open country, mountain, forest, and suburb with 210 to 410 images per category. We follow the experimental protocol as described in [17] wherein 1500 images are used for training whereas the remaining 2985 images are used for testing. The train/test split is determined randomly with the criterion that 100 images are selected for every scene category as train images and the remaining images are used as test images.

Table 5.5 shows the comparison of the proposed SKML features with popular learning methods. The LLC method [38] extracts a feature descriptor by using a locality constraint for projection to a local co-ordinate system. The DHFVC method [11] uses a hierarchical visual feature coding architecture based on restricted Boltzmann machines (RBM) for encoding of SIFT descriptors. A over-complete dictionary is learned by the D-KSVD algorithm [52] by integrating the classification error to the objective criterion whereas the LC-KSVD approach [15] adds a label consistency constraint combined with the classification and reconstruction error to form

Table 5.5 Comparison between the proposed method and other popular methods on the fifteen scene categories dataset

Method	Accuracy (%)
KSPM [17]	81.40
DHFVC [11]	86.40
LLC [38]	80.57
LaplacianSC [9]	89.70
D-KSVD [52]	89.10
LC-KSVD [15]	90.40
Hybrid-CNN [55]	91.59
SKML	**96.25**

a single objective function. Another popular sparse coding method is LaplacianSC [9] which preserves the locality of features by using a similarity preserving criterion based on Laplacian framework. The sparse coding methods D-KSVD, LC-KSVD and LaplacianSC achieves an accuracy of 89.10%, 90.40% and 89.70% respectively. The state-of-the-art deep learning method such as hybrid CNN [55] which is trained on a combination of training set of ImageNet-CNN and Places-CNN achieves a performance of 91.59%. The experimental results in Table 5.5 show that our proposed SKML method achieves higher performance of 96.25% compared to popular sparse coding and deep learning methods.

The confusion diagram for the fifteen scene categories dataset is shown in Fig. 5.8. The suburb category out of the fifteen scene categories achieves the best classification

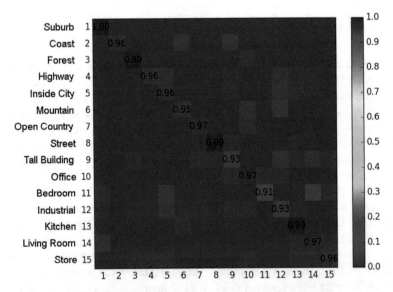

Fig. 5.8 The confusion matrix diagram of the 15 scene categories dataset using the SKML feature

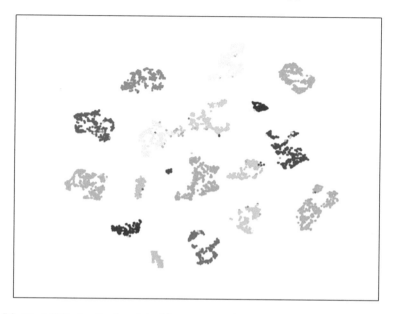

Fig. 5.9 The t-SNE visualization of the 15 scene categories dataset using SKML features

rate of 100%. The scene category with the lowest accuracy is the bedroom category with a classification rate of 91% as it has large confusions with the living room category. The living room scene category contains similar visual elements as the bedroom scene category resulting in high confusions between the two categories. The other scene categories that create confusion are tall building and industrial since both categories have some common visual semantics.

Figure 5.9 shows the t-SNE visualization for the fifteen scene categories dataset. The t-SNE method is a visualization technique used to fit high dimensional data to a plot using a non-linear dimensionality reduction technique to better understand the clusters of data of different categories in a dataset. It can be seen from Fig. 5.9 that our proposed SKML method improves the separability between clusters of different class. Another advantage of our proposed method is that it encourages better localization of data-points belonging to the same class resulting in better performance.

5.4.3   CalTech 101 Dataset

The Caltech 101 dataset [7] contains 9144 images of objects belonging to 101 categories. Every category has about 40 to 800 images and size of each image is roughly 300 X 200 pixels. The experimental protocol used for the CalTech 101 dataset is described in [38]. In particular, the training procedure involves five sets where each set contains 30, 25, 20, 15 and 10 train images per category respectively and for every

Table 5.6 Comparison between the proposed SKML method and other popular methods on the Caltech 101 dataset

Method	10	15	20	25	30
LLC [38]	59.77	65.43	67.74	70.16	73.44
SPM [17]	–	56.40	–	–	64.60
SVM-KNN [51]	55.80	59.10	62.00	–	66.20
SRC [41]	60.10	64.90	67.70	69.20	70.70
D-KSVD [52]	59.50	65.10	68.60	71.10	73.00
LC-KSVD [15]	63.10	67.70	70.50	72.30	73.60
CNN-M + Aug [4]	–	–	–	–	87.15
SKML	**82.47**	**84.46**	**85.35**	**86.61**	**87.95**

set, the test split contains the remaining images. In order to have a fair comparison with other methods, we report the performance as the average accuracy over all the categories.

The experimental results in Table 5.6 shows the detailed classification performance of the proposed SKML mathod and other popular learning methods for the CalTech 101 dataset. The SPM (spatial pyramid matching) method [17] divides an image to sub-regions and computes histogram over these sub-regions to form a spatial pyramid. The SVM-KNN method [51] finds the nearest neighbors of the query image and trains a local SVM based on the distance matrix computed on the nearest neighbors. The sparse coding method SRC [41] uses a sparse representation method computed by l_1 minimization and achieves classification accuracy of 70.70% for the set with training size 30 images per category. The deep learning method CNN-M + Aug [4] is similar to the architecture of ZFNet [50] but also incorporates additional augmentation techniques such as flipping and cropping to increase the training size. It can be seen from Table 5.6 that our proposed method achieves better performance compared to other learning methods. Another advantage of the SKML method is that no additional data augmentation techniques are required to improve the performance.

The t-SNE visualization for the CalTech 101 dataset is shown in Fig. 5.10. It can be seen that our proposed SKML method helps to increase the interclass separability between clusters having data-points belonging to different class categories as our method integrates a discriminative criterion to the objective function encouraging better clustering of data-points. Another advantage is that our method reduces the intraclass distance between data-points belonging to the same class in a cluster resulting in improved pattern recognition performance.

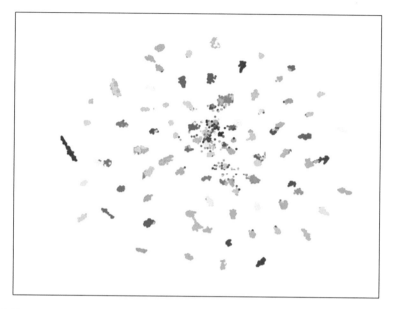

Fig. 5.10 The t-SNE visualization for the CalTech 101 dataset using SKML features

5.5 Conclusion

This chapter presents a sparse kernel manifold learner framework for different image classification applications. First, a new hybrid feature extraction step is performed by introducing D-FV and WS-FV features to capture different aspects of image and encode important discriminatory information. We then derive an innovative FFV feature by integrating the D-FV, WS-FV and SIFT-FV features. The FFV features are computed in eight different color spaces and fused to produce the novel FCFV feature. Finally, we propose a sparse kernel manifold learner (SKML) method by integrating a discriminative marginal Fisher criterion to the representation criterion to improve the classification performance. The SKML method aims to enhance the intraclass compactness and maximize the interclass separability constrained by the discriminative sparse objective function. Experimental results on different image classification datasets show the effectiveness of the proposed method.

References

1. Amiri, S.M., Nasiopoulos, P., Leung, V.C.M.: Non-negative sparse coding for human action recognition. In: 2012 19th IEEE International Conference on Image Processing, pp. 1421–1424 (2012)
2. Bosch, A., Zisserman, A., Munoz, X.: Representing shape with a spatial pyramid kernel. In: Proceedings of the 6th ACM International Conference on Image and Video Retrieval, CIVR '07, pp. 401–408 (2007)

3. Cai, D., He, X., Zhou, K., Han, J., Bao, H.: Locality sensitive discriminant analysis. In: Proceedings of the 20th International Joint Conference on Artifical Intelligence, IJCAI'07, pp. 708–713 (2007)
4. Chatfield, K., Simonyan, K., Vedaldi, A., Zisserman, A.: Return of the devil in the details: delving deep into convolutional nets. In: BMVC (2014)
5. Chen, J., Shan, S., He, C., Zhao, G., Pietikainen, M., Chen, X., Gao, W.: Wld: a robust local image descriptor. IEEE Trans. Pattern Anal. Mach. Intell. **32**(9), 1705–1720 (2010)
6. Chen, S., Liu, C.: Clustering-based discriminant analysis for eye detection. IEEE Trans. Image Processing **23**(4), 1629–1638 (2014)
7. Fei-Fei, L., Fergus, R., Perona, P.: Learning generative visual models from few training examples: an incremental bayesian approach tested on 101 object categories. In: CVPRW, pp. 178–178 (2004)
8. Fukunaga, K.: Introduction to Statistical Pattern Recognition. Academic Press Professional Inc, San Diego, CA, USA (1990)
9. Gao, S., Tsang, I.W.H., Chia, L.T.: Laplacian sparse coding, hypergraph laplacian sparse coding, and applications. IEEE Trans. Pattern Anal. Mach. Intell. **35**(1), 92–104 (2013)
10. Gao, Z., Liu, A., Zhang, H., Xu, G., Xue, Y.: Human action recognition based on sparse representation induced by l1/l2 regulations. In: 21st International Conference on Pattern Recognition (ICPR), pp. 1868–1871 (2012)
11. Goh, H., Thome, N., Cord, M., Lim, J.H.: Learning deep hierarchical visual feature coding. IEEE Trans. Neural Netw. Learn. Syst. **25**(12), 2212–2225 (2014)
12. Guo, Z., Zhang, D., Zhang, D.: A completed modeling of local binary pattern operator for texture classification. IEEE Trans. Image Process. **19**(6), 1657–1663 (2010)
13. He, X., Yan, S., Hu, Y., Niyogi, P., Zhang, H.J.: Face recognition using laplacianfaces. IEEE Trans. Pattern Anal. Mach. Intell. **27**(3), 328–340 (2005)
14. Jegou, H., Perronnin, F., Douze, M., Sanchez, J., Perez, P., Schmid, C.: Aggregating local image descriptors into compact codes. IEEE Trans. Pattern Anal. Mach. Intell. **34**(9), 1704–1716 (2012)
15. Jiang, Z., Lin, Z., Davis, L.S.: Label consistent k-svd: learning a discriminative dictionary for recognition. IEEE Trans. Pattern Anal. Mach. Intell. **35**(11), 2651–2664 (2013)
16. Khan, F., Beigpour, S., van de Weijer, J., Felsberg, M.: Painting-91: a large scale database for computational painting categorization. Mach. Vision Appl. **25**(6), 1385–1397 (2014)
17. Lazebnik, S., Schmid, C., Ponce, J.: Beyond bags of features: spatial pyramid matching for recognizing natural scene categories. In: Proceedings of the IEEE Conference on CVPR, vol. 2, pp. 2169–2178 (2006)
18. Li, Q., Schonfeld, D.: Multilinear discriminant analysis for higher-order tensor data classification. IEEE Trans. Pattern Anal. Mach. Intell. **36**(12), 2524–2537 (2014)
19. Liu, C.: Enhanced independent component analysis and its application to content based face image retrieval. Trans. Sys. Man Cyber. Part B **34**(2), 1117–1127 (2004). doi:10.1109/TSMCB. 2003.821449
20. Liu, C.: Gabor-based kernel pca with fractional power polynomial models for face recognition. IEEE Trans. Pattern Anal. Mach. Intell. **26**(5), 572–581 (2004)
21. Liu, C.: Extracting discriminative color features for face recognition. Pattern Recogn. Lett. **32**(14), 1796–1804 (2011)
22. Liu, C.: Discriminant analysis and similarity measure. Pattern Recogn. **47**(1), 359–367 (2014)
23. Liu, C., Wechsler, H.: Gabor feature based classification using the enhanced fisher linear discriminant model for face recognition. Trans. Img. Proc. **11**(4), 467–476 (2002). doi:10.1109/TIP.2002.999679
24. Lowe, D.: Distinctive image features from scale-invariant keypoints. Int. J. Comput. Vision **60**(2), 91–110 (2004)
25. Ojala, T., Pietikainen, M., Maenpaa, T.: Multiresolution gray-scale and rotation invariant texture classification with local binary patterns. IEEE Trans. Pattern Anal. Mach. Intell. **24**(7), 971–987 (2002)

26. Olshausen, B., Field, D.: Emergence of simple-cell receptive field properties by learning a sparse code for natural images. Nature **381**(6583), 607–609 (1996)
27. Olshausen, B., Field, D.: Sparse coding with an overcomplete basis set: a strategy employed by v1? Vision Res. **37**(23), 3311–3325 (1997)
28. Peng, K.C., Chen, T.: Cross-layer features in convolutional neural networks for generic classification tasks. In: 2015 IEEE International Conference on Image Processing (ICIP), pp. 3057–3061 (2015)
29. Peng, K.C., Chen, T.: A framework of extracting multi-scale features using multiple convolutional neural networks. In: 2015 IEEE International Conference on Multimedia and Expo (ICME), pp. 1–6 (2015)
30. Rathus, L.: Foundations of Art and Design. Wadsworth Cengage Learning, Boston, MA (2008)
31. van de Sande, K., Gevers, T., Snoek, C.: Evaluating color descriptors for object and scene recognition. IEEE Trans. Pattern Anal. Mach. Intell. **32**(9), 1582–1596 (2010)
32. Shechtman, E., Irani, M.: Matching local self-similarities across images and videos. In: CVPR, pp. 1–8 (2007)
33. Simonyan, K., Parkhi, O.M., Vedaldi, A., Zisserman, A.: Fisher Vector Faces in the Wild. In: BMVC (2013)
34. Sinha, A., Banerji, S., Liu, C.: New color gphog descriptors for object and scene image classification. Mach. Vis. Appl. **25**(2), 361–375 (2014)
35. Tola, E., Lepetit, V., Fua, P.: Daisy: an efficient dense descriptor applied to wide-baseline stereo. IEEE Trans. Pattern Anal. Mach. Intell. **32**(5), 815–830 (2010)
36. Wang, H., Yuan, C., Hu, W., Ling, H., Yang, W., Sun, C.: Action recognition using nonnegative action component representation and sparse basis selection. IEEE Trans. Image Process. **23**(2), 570–581 (2014)
37. Wang, J., Wonka, P., Ye, J.: Lasso screening rules via dual polytope projection. J. Mach. Learn. Res. **16**, 1063–1101 (2015)
38. Wang, J., Yang, J., Yu, K., Lv, F., Huang, T., Gong, Y.: Locality-constrained linear coding for image classification. In: Proceedings of the IEEE Conference on CVPR, pp. 3360–3367 (2010)
39. Wang, J., Zhou, J., Liu, J., Wonka, P., Ye, J.: A safe screening rule for sparse logistic regression. In: Ghahramani, Z., Welling, M., Cortes, C., Lawrence, N.D., Weinberger, K.Q. (eds.) Advances in Neural Information Processing Systems, vol. 27, pp. 1053–1061 (2014)
40. van de Weijer, J., Schmid, C., Verbeek, J., Larlus, D.: Learning color names for real-world applications. IEEE Trans. Image Process. **18**(7), 1512–1523 (2009)
41. Wright, J., Yang, A.Y., Ganesh, A., Sastry, S.S., Ma, Y.: Robust face recognition via sparse representation. IEEE Transa. Pattern Anal. Mach. Intell. **31**(2), 210–227 (2009)
42. Xiang, Z.J., Ramadge, P.J.: Fast lasso screening tests based on correlations. In: 2012 IEEE International Conference on Acoustics, Speech and Signal Processing (ICASSP), pp. 2137–2140 (2012)
43. Xiang, Z.J., Xu, H., Ramadge, P.J.: Learning sparse representations of high dimensional data on large scale dictionaries. In: Shawe-Taylor, J., Zemel, R.S., Bartlett, P.L., Pereira, F., Weinberger, K.Q. (eds.) Advances in Neural Information Processing Systems, pp. 900–908 (2011)
44. Xin, M., Zhang, H., Sun, M., Yuan, D.: Recurrent temporal sparse autoencoder for attention-based action recognition. In: 2016 International Joint Conference on Neural Networks (IJCNN), pp. 456–463 (2016)
45. Yan, S., Xu, D., Zhang, B., Zhang, H., Yang, Q., Lin, S.: Graph embedding and extensions: a general framework for dimensionality reduction. IEEE Trans. Pattern Anal. Mach. Intell. **29**(1), 40–51 (2007)
46. Yan, Y., Ricci, E., Subramanian, R., Liu, G., Sebe, N.: Multitask linear discriminant analysis for view invariant action recognition. IEEE Trans. Image Process. **23**(12), 5599–5611 (2014)
47. Yang, M., Zhang, L., Feng, X., Zhang, D.: Fisher discrimination dictionary learning for sparse representation. In: 2011 International Conference on Computer Vision, pp. 543–550 (2011)
48. Yang, M., Zhang, L., Feng, X., Zhang, D.: Sparse representation based fisher discrimination dictionary learning for image classification. Int. J. Comput. Vision **109**(3), 209–232 (2014)

49. Yuan, C., Hu, W., Tian, G., Yang, S., Wang, H.: Multi-task sparse learning with beta process prior for action recognition. In: 2013 IEEE Conference on Computer Vision and Pattern Recognition (CVPR), pp. 423–429 (2013)
50. Zeiler, M.D., Fergus, R.: Visualizing and understanding convolutional networks. In: European Conference on Computer Vision, pp. 818–833. Springer (2014)
51. Zhang, H., Berg, A.C., Maire, M., Malik, J.: Svm-knn: discriminative nearest neighbor classification for visual category recognition. In: Proceedings of the IEEE Conference on CVPR, vol. 2, pp. 2126–2136 (2006)
52. Zhang, Q., Li, B.: Discriminative k-svd for dictionary learning in face recognition. In: Proceedings of the IEEE Conference on CVPR, pp. 2691–2698 (2010)
53. Zhang, X., Chu, D., Tan, R.C.E.: Sparse uncorrelated linear discriminant analysis for under-sampled problems. IEEE Trans. Neural Netw. Learn. Syst. **27**(7), 1469–1485 (2016)
54. Zheng, J., Jiang, Z.: Learning view-invariant sparse representations for cross-view action recognition. In: 2013 IEEE International Conference on Computer Vision, pp. 3176–3183 (2013)
55. Zhou, B., Lapedriza, A., Xiao, J., Torralba, A., Oliva, A.: Learning deep features for scene recognition using places database. In: Proceedings of the NIPS, pp. 487–495 (2014)
56. Zhou, N., Shen, Y., Peng, J., Fan, J.: Learning inter-related visual dictionary for object recognition. In: 2012 IEEE Conference on Computer Vision and Pattern Recognition (CVPR), pp. 3490–3497 (2012)

Chapter 6
A New Efficient SVM (eSVM) with Applications to Accurate and Efficient Eye Search in Images

Shuo Chen and Chengjun Liu

Abstract This chapter presents an efficient Support Vector Machine (eSVM) for image search and video retrieval in general and accurate and efficient eye search in particular. Being an efficient and general learning and recognition method, the eSVM can be broadly applied to various tasks in intelligent image search and video retrieval. The eSVM first defines a Θ set that consists of the training samples on the wrong side of their margin derived from the conventional soft-margin SVM. The Θ set plays an important role in controlling the generalization performance of the eSVM. The eSVM then introduces only a single slack variable for all the training samples in the Θ set, and as a result, only a very small number of those samples in the Θ set become support vectors. The eSVM hence significantly reduces the number of support vectors and improves the computational efficiency without sacrificing the generalization performance. The optimization of the eSVM is implemented using a modified Sequential Minimal Optimization (SMO) algorithm to solve the large Quadratic Programming (QP) problem. A new eye localization method then applies the eSVM for accurate and efficient eye localization. In particular, the eye localization method consists of the eye candidate selection stage and the eye candidate validation stage. The selection stage selects the eye candidates from an image through a process of eye color distribution analysis in the YCbCr color space. The validation stage applies first 2D Haar wavelets for multi-scale image representation, then PCA for dimensionality reduction, and finally the eSVM for classification. Experiments on several diverse data sets show that the eSVM significantly improves the computational efficiency upon the conventional SVM while achieving comparable classification performance with the SVM. Furthermore, the eye localization results on the Face Recognition Grand Challenge (FRGC) database and the FERET database reveal that the proposed eye localization method achieves real-time eye detection speed and better eye detection performance than some recent eye detection methods.

S. Chen (✉)
The Neat Company, Philadelphia, PA 19103, USA
e-mail: sc77@njit.edu

C. Liu (✉)
New Jersey Institute of Technology, Newark, NJ 07102, USA
e-mail: chengjun.liu@njit.edu

© Springer International Publishing AG 2017
C. Liu (ed.), *Recent Advances in Intelligent Image Search and Video Retrieval*,
Intelligent Systems Reference Library 121, DOI 10.1007/978-3-319-52081-0_6

6.1 Introduction

Support Vector Machine (SVM) [37] has become a popular method in object detection and recognition [9–12, 25, 31]. When the detection or recognition problem becomes complex, the number of support vectors tends to increase, which subsequently leads to the increase of the model complexity. As a result, SVM becomes less efficient due to the expensive computation cost of its decision function, which involves an inner product of all the support vectors for the linear SVM and a kernel computation of all the support vectors for the kernel SVM. A number of simplified SVMs have been proposed to address the inefficiency problem of the conventional SVM. Burges [1] proposed a method, which computes an approximation to the decision function using a reduced set of support vectors and thus reduces the computation complexity of the conventional SVM by a factor of ten. This method was then applied to handwritten digits recognition [34] and face detection [33]. The authors in [15, 16] presented a Reduced Support Vector Machine (RSVM) as an alternative to the standard SVM for improving the computational efficiency [15, 16]. The RSVM generates a nonlinear kernel based on a separating surface (decision function) by solving a smaller optimization problem using a subset of training samples. The RSVM successfully reduces the model complexity and the computation cost. Other new SVM models include the υ-SVM [6], the simplified SVM [26], and the mirror classifier based SVM [4]. One drawback of these new SVMs is that they tend to reduce the classification accuracy when improving the computational efficiency.

We present in this paper an efficient Support Vector Machine (eSVM) that improves the computational efficiency of the conventional SVM without sacrificing the generalization performance. The eSVM first defines a Θ set that consists of the training samples on the wrong side of their margin derived from the conventional soft-margin SVM. The Θ set plays an important role in controlling the generalization performance of the eSVM. The eSVM then introduces only a single slack variable for all the training samples in the Θ set, and as a result, only a very small number of those samples in the Θ set become support vectors. In contrast, the conventional soft-margin SVM usually derives a large number of support vectors for the nonseparable case, because all the training samples on the wrong side of their margin (i.e., samples in the Θ set) become support vectors due to the introduction of the independent and different slack variables. The eSVM therefore significantly reduces the computation complexity of the conventional SVM. As the optimization problem of the eSVM is defined differently from that of the conventional SVM, we further present a modified Sequential Minimal Optimization (SMO) algorithm to solve the large convex quadratic programming problem defined in the eSVM. For experiments, we first compare the eSVM method with the SVM method using 18 data sets: a synthetic data set and 17 publicly accessible data sets used by other researchers [5, 8, 26, 29, 32]. We then compare the eSVM method with some other simplified SVM methods, such as the Reduced SVM (RSVM) [16], using 6 large-scale data sets [8]. Experimental results show that our eSVM method significantly improves the computational efficiency upon the conventional SVM and the RSVM while achieving comparable classification performance to or higher performance than the SVM and the RSVM.

We then present an accurate and efficient eye localization method using the eSVM. Accurate and efficient eye localization is important for building a fully automatic face recognition system [18–20], and the challenges for finding a robust solution to this problem have attracted much attention in the pattern recognition community [3, 7, 10, 13, 14, 25, 28, 39]. Example challenges in accurate and efficient eye localization include large variations in image illumination, skin color (white, yellow, and black), facial expression (eyes open, partially open, or closed), as well as scale and orientation. Additional challenges include eye occlusion caused by eye glasses or long hair, and the red eye effect due to the photographic effect. All these challenge factors increase the difficulty of accurate and efficient eye-center localization.

Our eye localization method consists of two stages: the eye candidate selection stage and the eye candidate validation stage. The selection stage rejects 99% of the pixels through an eye color distribution analysis in the YCbCr color space [35], while the remaining 1% of the pixels are further processed by the validation stage. The validation stage applies 2D Haar wavelets for multi-scale image representation, Principal Component Analysis (PCA) for dimensionality reduction, and the eSVM for classification to detect the center of the eye. We then evaluate our eSVM based eye localization method on the Face Recognition Grand Challenge (FRGC) version 2 database [27] and the FERET database [28]. The experimental results show that our proposed eye localization method achieves real-time eye detection speed and better eye detection performance than some recent eye detection methods.

6.2 An Efficient Support Vector Machine

We present in this section an efficient Support Vector Machine (eSVM), which significantly improves the computational efficiency upon the conventional SVM without sacrificing its generalization performance. Next, we first briefly review the conventional Support Vector Machine (SVM), followed by the analysis of the factors that cause the inefficiency problem (i.e., the large number of support vectors) of the conventional SVM.

6.2.1 The Conventional Soft-Margin Support Vector Machine

Let the training set be $\{(\mathbf{x}_1, y_1), (\mathbf{x}_2, y_2), \ldots, (\mathbf{x}_l, y_l)\}$, where $\mathbf{x}_i \in \mathbb{R}^n$, $y_i \in \{-1, 1\}$ indicate the two different classes, and l is the number of the training samples. When the training samples are linearly separable, the conventional SVM defines an optimal separating hyperplane, $\mathbf{w}^t \mathbf{x} + b = 0$, by minimizing the following functional:

$$\min_{\mathbf{w},b} \tfrac{1}{2}\mathbf{w}^t\mathbf{w},$$
$$subject\ to\ \ y_i(\mathbf{w}^t\mathbf{x}_i + b) \geq 1,\ \ i = 1, 2, \ldots, l. \tag{6.1}$$

When the training samples are not linearly separable, the conventional soft-margin SVM determines the soft-margin optimal hyperplane by minimizing the following functional:

$$\min_{\mathbf{w},b,\xi_i} \tfrac{1}{2}\mathbf{w}^t\mathbf{w} + C \sum_{i=1}^{l} \xi_i ,$$
$$subject\ to\ \ y_i(\mathbf{w}^t\mathbf{x}_i + b) \geq 1 - \xi_i,\ \ \ \xi_i \geq 0,\ \ i = 1, 2, \ldots, l. \tag{6.2}$$

where $\xi_i \geq 0$ are slack variables and $C > 0$ is a regularization parameter.

The Lagrangian theory and the Kuhn–Tucker theory are then applied to optimize the functional in Eq. 6.2 with inequality constraints [37]. The optimization process leads to the following quadratic convex programming problem:

$$\max_{\alpha} \sum_{i=1}^{l} \alpha_i - \tfrac{1}{2} \sum_{i,j=1}^{l} \alpha_i\alpha_j y_i y_j \mathbf{x}_i \mathbf{x}_j$$
$$subject\ to\ \ \sum_{i=1}^{l} y_i\alpha_i = 0,\ \ \ \ 0 \leq \alpha_i \leq C,\ \ i = 1, 2, \ldots, l \tag{6.3}$$

From the Lagrangian theory and the Kuhn–Tucker theory, we also have:

$$\mathbf{w} = \sum_{i=1}^{l} y_i\alpha_i\mathbf{x}_i = \sum_{i \in SV} y_i\alpha_i\mathbf{x}_i \tag{6.4}$$

where SV is the set of Support Vectors (SVs), which are the training samples with nonzero coefficients α_i. The decision function of the SVM is therefore derived as follows:

$$f(x) = sign(\mathbf{wx} + b) = sign(\sum_{i \in SV} y_i\alpha_i\mathbf{x}_i\mathbf{x} + b) \tag{6.5}$$

Equation 6.5 reveals that the computation of the decision function involves an inner product of all the support vectors. (Note that the computation of the decision function for the kernel SVM involves a kernel computation of all the support vectors.) Therefore, the computation cost of the decision function is proportional to the number of the support vectors. When the number of the support vectors is large, the computation cost of the inner product will become expensive and the computational efficiency of the conventional soft-margin SVM will be compromised. According to the Kuhn–Tucker theory, we have the following conditions for the conventional soft-margin SVM:

$$\alpha_i[y_i(\mathbf{w}^t\mathbf{x}_i + b) - 1 + \xi_i] = 0,\ \ i = 1, 2, \ldots, l. \tag{6.6}$$

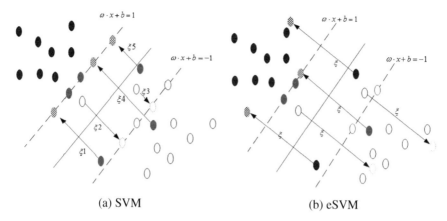

(a) SVM   (b) eSVM

Fig. 6.1 a Illustration of the conventional soft-margin SVM in the two dimensional space, where all five samples on the wrong side of their margin are pulled onto their boundaries to become support vectors. **b** Illustration of the eSVM in the two dimensional space, where only one of the five samples on the wrong side of their margin is pulled onto its boundary to become a support vector

Equation 6.6 shows that if $\alpha_i \neq 0$, then $y_i(\mathbf{w}^t\mathbf{x}_i + b) - 1 + \xi_i = 0$. Therefore, the training samples that satisfy $y_i(\mathbf{w}^t\mathbf{x}_i + b) - 1 + \xi_i = 0$ are support vectors for the conventional soft-margin SVM. The intuitive interpretation of the support vectors is that they are the training samples that lie on their boundaries or the samples pulled onto their boundaries by the slack variables ξ_i as shown in Fig. 6.1a. In fact, Fig. 6.1a shows that all the training samples on the wrong side of their margin become support vectors because of the slack variables, which pull the training samples onto their boundaries to make them support vectors. As a complex pattern classification problem often has a large number of the training samples on the wrong side of their margin, the number of support vectors becomes quite large, which leads to the inefficiency problem of the conventional soft-margin SVM.

6.2.2 Efficient Support Vector Machine

To address the inefficiency problem of the conventional soft-margin SVM, we present a new SVM, the efficient SVM (eSVM). As we discussed above, the conventional soft-margin SVM usually derives a large number of support vectors for the non-separable case, because all the training samples on the wrong side of their margin become support vectors as the slack variables pull these samples to their boundaries. The eSVM, however, reduces the number of support vectors significantly, because only a small number (can be as few as one) of the training samples on the wrong side of their margin are pulled to their boundaries to become support vectors. The eSVM first defines a Θ set that contains the training samples on the wrong side of their margin derived from the conventional soft-margin SVM. The Θ set plays an important role in controlling the generalization performance of the eSVM. The eSVM then introduces only a single slack variable for all the training samples in the Θ set, and as a result, only a very small number of those samples in the Θ set are pulled to their boundaries and become support vectors. As the number of support vectors reduced, the eSVM improve the computational efficiency upon the conventional soft-margin SVM.

Specifically, the eSVM optimizes the following functional:

$$
\min_{\mathbf{w},b,\xi} \tfrac{1}{2}\mathbf{w}^t\mathbf{w} + C\xi ,
$$
$$
subject\ to\ \ y_i(\mathbf{w}^t\mathbf{x}_i + b) \geq 1 ,\ i \in \Omega - \Theta \tag{6.7}
$$
$$
y_i(\mathbf{w}^t\mathbf{x}_i + b) \geq 1 - \xi ,\ i \in \Theta ,\ \xi \geq 0
$$

where Θ is the set of the training samples on the wrong side of their margin derived from the conventional soft-margin SVM, and Ω is the set of all the training samples. Compared with the conventional soft-margin SVM that defines the slack variables with different values, our eSVM specifies a fixed value for all the slack variables. The first inequality constraint in Eq. 6.7 ensures that the training samples on the right side of their margin in the conventional soft-margin SVM are still on the right side in the eSVM. The second inequality constraint in Eq. 6.7 ensures that only a small number of training samples in the Θ set becomes support vectors due to the introduction of the single slack variable that pulls most of the training samples in the Θ set beyond their boundary to the right side and thus become non-support vectors. The significance of the eSVM is to simulate the maximal margin classifier of the conventional SVM by using fewer support vectors.

Further analysis on the first inequality constraint in Eq. 6.7 reveals that this constraint makes the eSVM to maintain a similar maximal margin separating boundary with that of the conventional SVM. The definition of the separating boundary $\mathbf{w}^t\mathbf{x} + b = 0$ is associated with the definition of the maximal margin $\mathbf{w}^t\mathbf{x} + b = \pm 1$. Given the separating boundary $\mathbf{w}^t\mathbf{x} + b = 0$, the maximal margin is fixed to $\mathbf{w}^t\mathbf{x} + b = \pm 1$, and vice versa. The first inequality constraint in Eq. 6.7 ensures that the training samples on the right side of their margin in the conventional soft-margin SVM are still on the right side in the eSVM. If the eSVM derives a separating boundary that is significantly different from the one derived by the conventional SVM, this constraint will not be satisfied. As the eSVM derives a similar separating boundary with the conventional SVM, the maximal margin of the eSVM is thus similar with that of the conventional SVM as well – neither degrade nor upgrade much from that of the SVM. Consequently, the eSVM inherits the advantage of generalization performance of the conventional SVM, and has comparable classification performance with the SVM.

The second inequality constraint in Eq. 6.7 plays the significant role of reducing the number of support vectors. This constraint ensures that only a small number of training samples in the Θ set becomes support vectors due to the introduction of the single slack variable that pulls most of the training samples in the Θ set beyond their boundary to the right side and thus become non-support vectors. It is possible that support vectors lying on their boundaries for the eSVM are a little bit more than those for the SVM. However, majority of support vectors for the SVM come from the samples on the wrong side of their margin (i.e., the Θ set). This is because the chance that samples happen to fall on their boundaries is significantly lower than the chance that samples fall onto the right side or wrong side of their margin. Therefore, even though support vectors on their boundaries for the eSVM may increase a little bit upon those for the SVM, the total number of support vectors for the eSVM is still significantly less than that of the SVM.

As the optimization problem of the eSVM defined in Eq. 6.7 is different from that of the conventional SVM, its corresponding dual mathematical problem after applying the Lagrangian theory and the Kuhn–Tucker theory is also different from the one derived in the conventional soft-margin SVM. In particular, let $\alpha_1, \alpha_2, \ldots, \alpha_l \geq 0$ and $\mu \geq 0$ be the Lagrange multipliers, the primal Lagrange functional is defined as follows:

$$
\begin{aligned}
\mathscr{F}(\mathbf{w}, b, \xi, \alpha_i, \mu) = {} & \tfrac{1}{2}\mathbf{w}^t\mathbf{w} + C\xi - \sum_{i\in\Omega-\Theta} \alpha_i[y_i(\mathbf{w}^t\mathbf{x}_i - b) - 1] \\
& - \sum_{i\in\Theta} \alpha_i[y_i(\mathbf{w}^t\mathbf{x}_i - b) + \xi - 1] - \mu\xi \\
= {} & \tfrac{1}{2}\mathbf{w}^t\mathbf{w} + (C - \sum_{i\in\Theta}\alpha_i - \mu)\xi - \sum_{i\in\Omega}\alpha_i[y_i(\mathbf{w}^t\mathbf{x}_i - b) - 1]
\end{aligned}
$$

(6.8)

Next, we maximize the primal Lagrange functional $\mathscr{F}(\mathbf{w}, b, \xi, \alpha_i, \mu)$ with respect to $\mathbf{w}$, b, and ξ as follows:

$$
\frac{\partial\mathscr{F}}{\partial\mathbf{w}} = \mathbf{w} - \sum_{i\in\Omega}\alpha_i y_i\mathbf{x}_i = 0 \;\Rightarrow\; \mathbf{w} = \sum_{i\in\Omega}\alpha_i y_i\mathbf{x}_i \tag{6.9}
$$

$$
\frac{\partial\mathscr{F}}{\partial b} = \sum_{i\in\Omega}\alpha_i y_i = 0 \tag{6.10}
$$

$$
\frac{\partial\mathscr{F}}{\partial\xi} = C - \sum_{i\in\Theta}\alpha_i - \mu = 0 \tag{6.11}
$$

Then, we derive a convex quadratic programming model by substituting Eqs. 6.9, 6.10, and 6.11 into Eq. 6.8 as follows:

$$
\max_{\alpha} \sum_{i\in\Omega}\alpha_i - \tfrac{1}{2}\sum_{i,j\in\Omega}\alpha_i\alpha_j y_i y_j\mathbf{x}_i\mathbf{x}_j
$$
$$
subject\ to\ \sum_{i\in\Omega} y_i\alpha_i = 0, \quad \left(\sum_{i\in\Theta}\alpha_i\right) \leq C, \quad \alpha_i \geq 0,\ i\in\Omega \tag{6.12}
$$

Furthermore, we have the following constraints from the Kuhn–Tucker theory:

$$
\begin{aligned}
\alpha_i[y_i(\mathbf{w}^t\mathbf{x} + b) - 1] = 0,\ i\in\Omega-\Theta \\
\alpha_i[y_i(\mathbf{w}^t\mathbf{x} + b) - 1 + \xi] = 0,\ i\in\Theta
\end{aligned} \tag{6.13}
$$

Equation 6.13 shows that if $\alpha_i \neq 0$, then either $y_i(\mathbf{w}^t\mathbf{x} + b) - 1 = 0, i \in \Omega - \Theta$ or $y_i(\mathbf{w}^t\mathbf{x} + b) - 1 + \xi = 0, i \in \Theta$. The training samples that satisfy either $y_i(\mathbf{w}^t\mathbf{x} + b) - 1 = 0, i \in \Omega - \Theta$ or $y_i(\mathbf{w}^t\mathbf{x} + b) - 1 + \xi = 0, i \in \Theta$ are support vectors for the eSVM. Therefore, the intuitive interpretation of the support vectors is the training samples that lie on their boundaries or the samples pulled onto their boundaries by the slack variable ξ as shown in Fig. 6.1b. As all the slack variables in the eSVM have the same value, Fig. 6.1b also reveals that only a small number (can be as few as one) of the training samples on the wrong side of their margin (i.e., samples in the Θ set) are pulled onto their boundaries to become support vectors, while the others are not support vectors because they are pulled to the right side of their margin but do not fall onto the boundaries.

6.2.3 Modified Sequential Minimal Optimization Algorithm

Both the SVM and the eSVM training procedure require a solution to a very large Quadratic Programming (QP) optimization problem. The Sequential Minimal Optimization (SMO) algorithm [30] is a popular and efficient tool to solve the large QP problem defined in the SVM. As the mathematical model in Eq. 6.12 of the eSVM is defined different from that of the conventional SVM, we present in this section a modified SMO algorithm for the eSVM to solve the convex QP problem in Eq. 6.12.

The modified SMO algorithm, as the SMO does, breaks a large QP problem into a series of the smallest possible optimization problems, which can be solved analytically without resorting to a time-consuming iterative process. The modified SMO algorithm consists of two major steps: an analytical solution to the smallest QP problem with two Lagrange multipliers, and a heuristic approach for choosing which two multipliers to optimize. Next, we present these two steps in details.

6.2.3.1 An Analytic Solution to the Smallest QP Problem

Let α_s and α_t, for one of the smallest QP problems, be the two Lagrange multipliers to be optimized while the other α_i's are fixed. First, the modified SMO algorithm, as the SMO does, derives the unconstrained maximum value for α_s and α_t:

$$\alpha_t^{new} = \alpha_t^{old} + \frac{y_t(g_s^{old} - g_t^{old})}{\eta}$$
$$\alpha_s^{new} = \gamma - \Delta\alpha_t^{new} \tag{6.14}$$

where $\Delta = y_s y_t$, $\gamma = \alpha_s^{old} + \Delta\alpha_t^{old}$, $\eta = 2\mathbf{x}_s\mathbf{x}_t - \mathbf{x}_s\mathbf{x}_s - \mathbf{x}_t\mathbf{x}_t$, $g_s^{old} = y_s - \mathbf{x}_s^t\mathbf{w}^{old}$, and $g_t^{old} = y_s - \mathbf{x}_t^t\mathbf{w}^{old}$. Note that for the initialization step, α^{old} can be set to 0.

Then, the two Lagrange multipliers α_s^{new} and α_t^{new} should be checked if they satisfy the inequality constraints in Eq. 6.12:

$$\alpha_i \geq 0, i \in \Omega, \left(\sum_{i \in \Theta} \alpha_i \right) \leq C$$

If the two Lagrange multipliers α_s^{new} and α_t^{new} in Eq. 6.14 do not satisfy the above inequality constraints, their values need to be adjusted. Depending on the values of α_s^{new}, α_t^{new}, and Δ, there are several cases to consider:

1. If $\Delta = 1$, then $\alpha_s^{new} + \alpha_t^{new} = \gamma$

   a. If $\alpha_s^{new}(\alpha_t^{new}) < 0$, then $\alpha_s^{new}(\alpha_t^{new}) = 0$, $\alpha_t^{new}(\alpha_s^{new}) = \gamma$;
   b. If $s \in MV$, $t \notin MV$, and $\alpha_s^{new} > C - \sum_{i \in MV, i \neq s} \alpha_i^{old}$, then
      $\alpha_s^{new} = C - \sum_{i \in MV, i \neq s} \alpha_i^{old}$, $\alpha_t^{new} = \gamma - \alpha_s^{new}$;
   c. If $s \notin MV$, $t \in MV$, and $\alpha_t^{new} > C - \sum_{i \in MV, i \neq t} \alpha_i^{old}$, then
      $\alpha_t^{new} = C - \sum_{i \in MV, i \neq t} \alpha_i^{old}$, $\alpha_s^{new} = \gamma - \alpha_t^{new}$;
   d. If $s \in MV$ and $t \in MV$, since $\alpha_s^{new} + \alpha_t^{new} = \alpha_s^{old} + \alpha_t^{old}$, there is no effect on $\sum_{i \in MV} \alpha_i$. So α_s^{new} and α_t^{new} don't need to adjust;
   e. If $s \notin MV$ and $t \notin MV$, since both α_s^{new} and α_t^{new} would not affect $\sum_{i \in MV} \alpha_i$, they don't need to adjust.

2. If $\Delta = -1$, then $\alpha_s^{new} - \alpha_t^{new} = \gamma$

   a. If $\alpha_s^{new} < 0$, then $\alpha_s^{new} = 0$, $\alpha_t^{new} = -\gamma$;
   b. If $\alpha_t^{new} < 0$, then $\alpha_t^{new} = 0$, $\alpha_s^{new} = \gamma$;
   c. If $s \in MV$, $t \notin MV$, and $\alpha_s^{new} > C - \sum_{i \in MV, i \neq s} \alpha_i^{old}$, then
      $\alpha_s^{new} = C - \sum_{i \in MV, i \neq s} \alpha_i^{old}$, $\alpha_t^{new} = \alpha_s^{new} - \gamma$;
   d. If $s \notin MV$, $t \in MV$, and $\alpha_t^{new} > C - \sum_{i \in MV, i \neq t} \alpha_i^{old}$, then
      $\alpha_t^{new} = C - \sum_{i \in MV, i \neq t} \alpha_i^{old}$, $\alpha_s^{new} = \alpha_t^{new} + \gamma$;
   e. If $s \in MV$, $t \in MV$, and $\alpha_s^{new} + \alpha_t^{new} > C - \sum_{i \in MV, i \neq s, t} \alpha_i^{old}$, then
      $\alpha_s^{new} = \frac{1}{2}(C - \sum_{i \in MV, i \neq s, t} \alpha_i^{old} + \gamma)$, $\alpha_t^{new} = \frac{1}{2}(C - \sum_{i \in MV, i \neq s, t} \alpha_i^{old} - \gamma)$;
   f. If $s \notin MV$ and $t \notin MV$, they don't need to adjust.

6.2.3.2 A Heuristic Approach for Choosing Two Multipliers

The SMO algorithm [30] applies some independent heuristics to choose which two Lagrange multipliers to optimize jointly at every step. The heuristics process is implemented by means of two loops: the outer loop selects the first α_i that violates the Kuhn–Tucker conditions, while the inner loop selects the second α_i that maximizes $|E_2 - E_1|$, where E_i is the prediction error on the ith training sample. Complexity analysis reveals that the conventional SMO algorithm, at every step of the optimization process, takes $O(l^2)$ to choose the Lagrange multipliers, where l is the number of the training samples. This process is time-consuming if the number of the training samples is very large.

The modified SMO algorithm, by contrast, applies an improved heuristic process for choosing multipliers. This process chooses the two α_i's simultaneously: each time the two α_i's that violate the Kuhn–Tucker conditions most seriously are chosen. The new heuristic process thus takes $O(l)$ to choose the Lagrange multipliers at every step.

In order to determine the pair of α_i's that violate the Kuhn–Tucker conditions most seriously, the Kuhn–Tucker conditions of the eSVM should be further analyzed. From Eq. 6.13, the Kuhn–Tucker conditions of the eSVM can be decomposed as follows:

1. When $i \in V - MV$

    a. If $\alpha_i > 0$ and $y_i = 1$, then $b = y_i - \omega^t \phi(x_i)$;
    b. If $\alpha_i > 0$ and $y_i = -1$, then $b = y_i - \omega^t \phi(x_i)$;
    c. If $\alpha_i = 0$ and $y_i = 1$, then $b \geq y_i - \omega^t \phi(x_i)$;
    d. If $\alpha_i = 0$ and $y_i = -1$, then $b \leq y_i - \omega^t \phi(x_i)$.

2. When $i \in MV$

    a. If $\alpha_i > 0$ and $y_i = 1$, then $b \leq y_i - \omega^t \phi(x_i)$;
    b. If $\alpha_i > 0$ and $y_i = -1$, then $b \geq y_i - \omega^t \phi(x_i)$;
    c. If $\sum_{j \in MV} \alpha_j < C$ and $y_i = 1$, then $b \geq y_i - \omega^t \phi(x_i)$;
    d. If $\sum_{j \in MV} \alpha_j < C$ and $y_i = -1$, then $b \leq y_i - \omega^t \phi(x_i)$.

As a result, the pair of α_i's that violate the Kuhn–Tucker conditions most seriously can be determined as follows:

$$
\begin{aligned}
s = argmax(&\{y_i - \mathbf{w}^t \mathbf{x}_i | \alpha_i \geq 0, y_i = 1, i \in \Omega - \Theta\}, \\
&\{y_i - \mathbf{w}^t \mathbf{x}_i | \alpha_i > 0, y_i = -1, i \in \Omega - \Theta\}, \\
&\{y_i - \mathbf{w}^t \mathbf{x}_i | \sum_{j \in \Theta} \alpha_j < C, y_i = 1, i \in \Theta\}, \\
&\{y_i - \mathbf{w}^t \mathbf{x}_i | \alpha_i > 0, y_i = -1, i \in \Theta\}). \\
t = argmin(&\{y_i - \mathbf{w}^t \mathbf{x}_i | \alpha_i > 0, y_i = 1, i \in \Omega - \Theta\}, \\
&\{y_i - \mathbf{w}^t \mathbf{x}_i | \alpha_i \geq 0, y_i = -1, i \in \Omega - \Theta\}, \\
&\{y_i - \mathbf{w}^t \mathbf{x}_i | \sum_{j \in \Theta} \alpha_j < C, y_i = -1, i \in \Theta\}, \\
&\{y_i - \mathbf{w}^t \mathbf{x}_i | \alpha_i > 0, y_i = 1, i \in \Theta\}).
\end{aligned}
\tag{6.15}
$$

6.3 Accurate and Efficient Eye Localization Using eSVM

We present an accurate and efficient eye localization method in this section by applying the eSVM together with color information and 2D Haar wavelets. Figure 6.2 shows the architecture of the proposed eye localization system. First, we apply the Bayesian Discriminating Features(BDF) method [17] to detect a face from an image

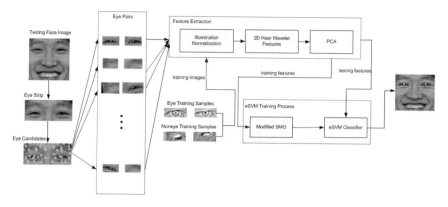

Fig. 6.2   System architecture of our eye localization method

and normalize the detected face to a predefined size (128×128 in our experiments). Second, we use some geometric constraints to extract an eye strip from the upper portion of the detected face (the size of the eye strip is 55×128 in our experiments). Third, we propose a new eye localization method that consists of two stages: the eye candidate selection stage and the eye candidate validation stage. Specifically, the selection stage rejects 99% of the pixels through an eye color distribution analysis in the YCbCr color space [35], while the remaining 1% of the pixels are further processed by the validation stage. The validation stage applies illumination normalization, 2D Haar wavelets for multi-scale image representation, PCA for dimensionality reduction, and the eSVM for classification, to detect the center of the eye. Next, we explain in detail the eye candidate selection and validation stages, respectively.

6.3.1   The Eye Candidate Selection Stage

The conventional sliding window based eye detection methods exhaustively classify all the pixels in an image from left to right, top to bottom to locate the eyes. The excessive number of the pixels over an image significantly slows down the classifier-based eye detection methods. We therefore propose a novel eye candidate selection stage to first dramatically reduce the number of eye candidates, which will be further validated by the classifier-based methods.

Specifically, the eye candidates are chosen through an eye color distribution analysis in the YCbCr color space [35]. In the YCbCr color space, the RGB components are separated into luminance (Y), chrominance blue (Cb), and chrominance red (Cr) [35]. In our research, we observe that in the eye region, especially around the pupil center, pixels are more likely to have higher values in chrominance blue (Cb) and lower values in chrominance red (Cr) when compared with those pixels in the skin region. We also observe that the luminance (Y) of the pixels in the eye region is

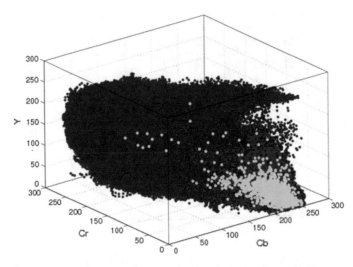

Fig. 6.3 The eye-tone distribution in the YCbCr color space. The skin pixels are represented in *red*, the eye region pixels are in *blue*, and the pupil-center pixels are in *green*

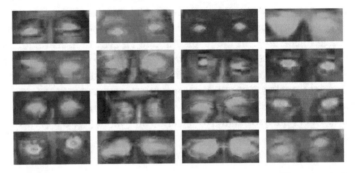

Fig. 6.4 Example eye strip images in the YCbCr color space, where Y is represented in *red*, Cb in *green*, and Cr in *blue*

much darker than those in the skin region. To illustrate these findings, we manually collect some random skin patches, eye regions, as well as pupil centers from 600 face images. The number of pixels randomly chosen from the skin patches and the eye regions are 4,078,800 pixels and 145,200 pixels, respectively. And the number of pixels corresponding to the pupil centers is 1,200 pixels as there are two eyes in each face image. Figure 6.3 shows the eye-tone distribution in the YCbCr color space, where the skin pixels are represented in red, the eye region pixels are in blue, and the pupil-center pixels are in green. Figure 6.3 reveals that the eye-centers, which are represented by green, are clustered in the corner with higher Cb values but lower Cr and Y values. Figure 6.4 shows some eye strip examples in the YCbCr color space, where Y is represented in red, Cb in green, and Cr in blue. One can see from Fig. 6.4 that the eye regions tend to have high green values and low blue values.

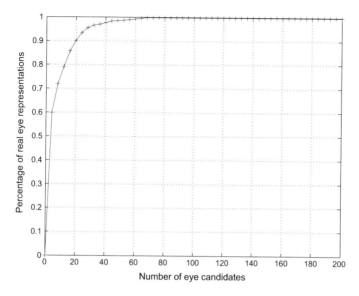

Fig. 6.5 The percentage of the real eye representations as the number of the selected eye candidates varies

Motivated by these findings, we present a new method for eye candidate selection. The idea of the eye candidate selection method is to define a weight for each pixel based on its Y, Cb, and Cr values and rank the pixels according to their weights. In particular, the weight of pixel (i, j) is defined as follows:

$$weight(i, j) = \sum_{i-2, j-2}^{i+2, j+2} [Cb(i, j) + (255 - Cr(i, j)) + (255 - Y(i, j))] \quad (6.16)$$

The first K pixels with maximum weights are therefore considered as the eye candidates. In Fig. 6.5, we randomly select 2,000 eye strip images from the FRGC database and the FERET database to evaluate the performance of our eye candidate selection method. The horizontal axis indicates the number of selected eye candidates (i.e., K), and the vertical axis indicates the percentage of the real eye representations. Please note that the real eye is considered represented if any of those eye candidates is within five pixels from the ground truth. Figure 6.5 shows that only 60 ($K = 60$) candidates per image through the eye color distribution analysis, which account for just 0.85% of the pixels over an image, can represent over 99% of the real eye locations on average. As a result, the significance of the eye candidate selection stage is that more than 99% ($1-0.85\%$) of the pixels over an image are rejected in this stage whereas only the remaining 1% of the pixels are further processed by the classifier-based validation stage. In comparison with the conventional sliding window method, the candidate selection stage dramatically reduces the number of eye candidates that will be validated by the classifier-based methods and hence significantly improves

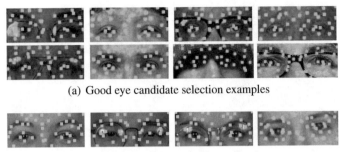

(a) Good eye candidate selection examples

(b) Bad eye candidate selection examples

Fig. 6.6 Examples of good (**a**) and bad (**b**) eye candidate selection results

the efficiency of the eye localization system. Figure 6.6 shows some examples of good and bad eye candidate selection results on the FRGC database and the FERET database. Please note that the result is considered bad if there is no pixel within five pixel distance from the ground truth chosen by the candidate selection method.

Another advantage of our eye candidate selection stage is that it can further improve the efficiency and accuracy of the eye localization system by considering the left and right eye candidates in pair to address the problem of scaling and rotation. Although all the training eye images are rotated upright and normalized to a fixed size, the testing images may vary in terms of size and orientation. Even though the traditional methods try to overcome these difficulties by searching a number of predefined scales and orientations, they are either time consuming or the incremental steps too large to cover the continuous scaling and rotation values well. Our eye candidate selection method solves the eye scaling and orientation problem by introducing an eye pair selection scheme that considers the eye candidates in pairs. In particular, all the eye candidates are divided into the left and the right eye candidates according to their relative positions in the eye strip. The left and right eye candidates are then formed in pairs, and the distance and angle of the binocular line of each eye pair is used to normalize and rotate the eye candidates to the predefined size and the upright orientation, respectively. Figure 6.7 shows an example of our eye pair selection scheme. The eye pair marked by the dark blue rectangles is considered the most suitable scale and orientation. Note that some eye pairs can be removed if the binocular distance is too small or too large, or the angle is too large.

Fig. 6.7 Example of our eye
pair selection scheme

6.3.2   The Eye Candidate Validation Stage

So far, the eye candidate selection stage selects a small number of eye candidates from each image. Our next stage will validate these eye candidates to find the real eyes from them. Specifically, the eye candidate validation stage applies illumination normalization, 2D Haar wavelets for multi-scale image representation, PCA for dimensionality reduction, and the eSVM for detection of the center of the eye. First, illumination normalization improves the quality of the eye strip image through Gamma correction, Difference of Gaussian (DoG) filtering, and contrast equalization [36].

Second, 2D Haar wavelets are then applied to extract the features for multi-scale image representation. The 2D Haar wavelet feature is defined as the projection of an image onto the 2D Haar basis functions [2]. The attractive characteristics of the 2D Haar basis functions enhance local contrast and facilitate feature extraction in many target detection problems, such as eye detection, where dark pupil is in the center of colored iris that is surrounded by white sclera. The 2D Haar basis functions can be generated from the one dimensional Haar scaling and wavelet functions. The Haar scaling function $\phi(x)$ may be defined as follows [2, 23]:

$$\phi(x) = \begin{cases} 1 & 0 \leq x < 1 \\ 0 & \text{otherwise} \end{cases} \tag{6.17}$$

A family of functions can be generated from the basic scaling function by scaling and translation [2, 23]:

$$\phi_{i,j}(x) = 2^{i/2}\phi(2^i x - j) \tag{6.18}$$

As a result, the scaling functions $\phi_{i,j}(x)$ can span the vector spaces V^i, which are nested: $V^0 \subset V^1 \subset V^2 \subset \cdots$ [23]. The Haar wavelet function $\psi(x)$ may be defined as follows [2, 23]:

$$\psi(x) = \begin{cases} 1 & 0 \leq x < 1/2 \\ -1 & 1/2 \leq x < 1 \\ 0 & \text{otherwise} \end{cases} \tag{6.19}$$

The Haar wavelets are generated from the mother wavelet by scaling and translation [2, 23]:

$$\psi_{i,j}(x) = 2^{i/2}\psi(2^i x - j) \tag{6.20}$$

The Haar wavelets $\psi_{i,j}(x)$ span the vector space W^i, which is the orthogonal complement of V^i in V^{i+1}: $V^{i+1} = V^i \oplus W^i$ [2, 23]. The 2D Haar basis functions are the tensor product of the one dimensional scaling and wavelet functions [23]. For example, for V^5, where $V^5 = V^0 \oplus W^0 \oplus W^1 \oplus W^2 \oplus W^3 \oplus W^4$, the 2D Haar basis consists of 1, 024 basis functions.

Third, Principal Component Analysis (PCA), which is the optimal method for feature representation [21], is applied to the Haar wavelet features for dimensionality reduction. In particular, let $Y \in \mathbb{R}^N$ represent the 2D Haar wavelet features. The covariance matrix of Y is defined as $\sum_Y = \varepsilon\{[Y - \varepsilon(Y)][Y - \varepsilon(Y)]^t\}$, where $\varepsilon(\cdot)$ is the expectation operator and $\sum_Y \in \mathbb{R}^{N \times N}$. PCA of a random vector Y factorizes the covariance matrix $\sum_Y$ into the following form: $\sum_Y = \Phi \Lambda \Phi$, where $\Phi = [\phi_1 \phi_2 \cdots \phi_N]$ is an orthogonal eigenvector matrix, and $\Lambda = diag\{\lambda_1, \lambda_2, \ldots, \lambda_N\}$ is a diagonal eigenvalue matrix with the diagonal elements in decreasing order $(\lambda_1 \geq \lambda_2 \geq \ldots \geq \lambda_N)$. An important application of PCA is dimensionality reduction: $Z = P^t Y$, where $P = [\phi_1 \phi_2 \ldots \phi_m]$, $m < N$ and $P \in \mathbb{R}^{N \times m}$.

Finally, the eSVM applies the PCA features to validate whether an eye candidate indeed represents the center of an eye. Note that when the eSVM is used for classification, the sign of its decision function only determines the class membership of the samples. It is reasonable that a number of candidates around the eye center will be classified into the eye category. In order to determine the final eye center location, the eSVM decision values, instead of the signs of the decision function, are used to first select Q eye candidates with bigger decision values. After the Q candidates are selected, following steps are introduced: first, for each eye candidate, consider an $n \times n$ square centered at this eye candidate; second, compute the summation of the eSVM decision values of all the eye candidates within this $n \times n$ square, and assign this summation to the eye candidate; finally, select the eye candidate with the highest summation as the center of the eye. Note that Q is determined empirically, and we will discuss in the next section the choice of Q and its effect on the performance of the eSVM classifier.

6.4 Experiments

We now evaluate our proposed eSVM method and its application to accurate and efficient eye localization. In particular, we first fully evaluate the eSVM method on several data sets that are from different classification problems and widely used by other SVM researchers. Three experiments are performed to evaluate both the classification and the efficiency performance of the eSVM. The first experiment runs on a synthetic data set [5, 26] to give an intuitive view in the two dimensional space of the eSVM in comparison with the SVM. The second experiment runs on 17 data sets from the UCI Adult benchmark collection and the Web Classification collection [8, 29, 32] to compare the performance of the eSVM with the SVM. The third experiment runs on 6 large-scale data sets [8] to compare the eSVM with other simplified SVM methods, such as the Reduced SVM (RSVM) [16]. Experimental results show that our eSVM method significantly improves the computational efficiency upon the conventional SVM and the RSVM while achieving comparable classification performance to or higher performance than the SVM and the RSVM. We then evaluate our eSVM based eye localization method on the Face Recognition Grand Challenge (FRGC) version 2 database [27] and the FERET database [28]. The experimental results show that our proposed eye localization method achieves real-time eye detection speed and better eye detection performance than some recent eye detection methods.

6.4.1 Performance Evaluation of the eSVM Method

We first evaluate the performance of the eSVM on a synthetic data set, the *Ripley* data set, which is also used in [5, 26]. The advantage of applying this data set comes from the intuitive visualization of the experimental results in the two dimensional space where the data resides, as the data has only two attributes. The *Ripley* data set defines a two-class non-separable problem, and the number of training and testing samples is 250 and 1,000, respectively. In our experiments, two runs are performed with different kernels and parameter settings – one run applies a linear kernel with the regularizing parameter $C = 100$, while the other run applies an RBF kernel $K(x_i, x_j) = e^{-r\|x_i - x_j\|^2}$ with the regularizing parameter $C = 100$ and power $r = 1$. Note that the parameter settings are the same as those in [5, 26].

Figure 6.8 plots the training samples, the support vectors, and the separating boundaries of the conventional SVM and the eSVM with the linear and RBF kernels, respectively. In particular, Fig. 6.8 shows that the proposed eSVM has similar separating boundaries to the conventional SVM using either linear or RBF kernels. The significance of this finding reveals that both eSVM and the conventional SVM have similar generalization performance. Another finding reveals that the number of support vectors for the eSVM is much smaller than that for the conventional SVM as shown in Fig. 6.8. Specifically, Fig. 6.8a, c show that the support vectors for the conventional SVM, which are represented by red crosses, are mostly the samples on the wrong side of their margin. The number of support vectors thus is large due to the fact that many training samples are on the wrong side of their margin for the non-separable problem. In contrast, for the eSVM, only a small number of the training samples on the wrong side of their margin, as shown in Fig. 6.8b, d, become support vectors. The number of support vectors for the eSVM is thus much smaller than that for the conventional SVM.

Table 6.1 shows the comparison of the SVM and the eSVM on the number of the support vectors, the slope and the y-intercept of the separating boundaries using the linear kernel, as well as the classification rate and running time for the testing data set. In particular, the eSVM reduces the number of support vectors by 88.76 and 75.64%, when compared with the conventional SVM using the linear and RBF kernels, respectively. Consequently, the running time of the eSVM is also reduced compared with the SVM when same kernel is applied. The similar slope and the y-intercept values of the separating boundaries between the eSVM and the conventional SVM using a linear kernel indicate that they define similar separating boundaries, and hence have comparable generalization performance. Specifically, the classification rate of the eSVM on the testing data set is the same with that of the SVM using the

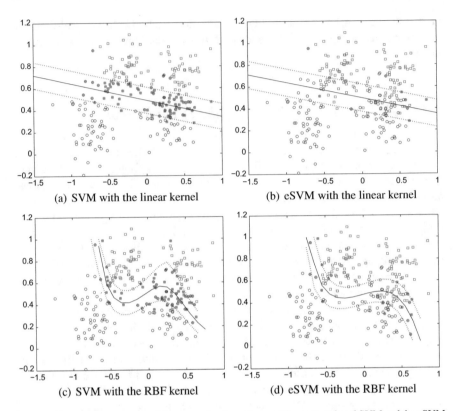

(a) SVM with the linear kernel

(b) eSVM with the linear kernel

(c) SVM with the RBF kernel

(d) eSVM with the RBF kernel

Fig. 6.8 The support vectors and the separating boundaries of the conventional SVM and the eSVM with the linear and RBF kernels on the *Ripley* data set, respectively. The *dashed lines*/curves depict the ± 1 margins around the separating boundary

Table 6.1 Result comparisons between the conventional SVM and the eSVM with the linear and RBF kernels

Method	#SV	Slope	y-intercept	Rate performance	Time (ms)
SVM – linear	89	−0.153	0.492	89.7 (897/1000)	14.29
eSVM – linear	10	−0.149	0.496	89.7 (897/1000)	6.42
SVM – RBF	78	–	–	89.7 (897/1000)	37.41
eSVM – RBF	19	–	–	90.2 (902/1000)	11.52

linear kernel, but the classification rate of the eSVM is 0.5% higher than that of the SVM when using the RBF kernel.

We then compare the performance of the eSVM and the SVM for accuracy and efficiency using 17 publicly available data sets – 9 from the UCI Adult benchmark collection and 8 from the Web Classification Collection [8, 29, 32]. The UCI Adult benchmark collection, also known as "Census Income" data set, is designed to predict whether a household has an income greater than $50,000. Each record in the UCI Adult collection contains 123 features. The Web Classification collection is used for text categorization problem. It collects the keywords from the web page as the attributes and classifies whether a web page belongs to a category or not. Each record in the Web collection contains 300 features extracted from a web page.

The parameters of the conventional SVM are set the same as those in [29], which are chosen to optimize accuracy on a validation set as done in [29]. Specifically, only the RBF kernel $K(x_i, x_j) = e^{-r\|x_i - x_j\|^2}$ is used. For the Adult data sets, the regularizing parameter C is set to 1 and the power r of the RBF kernel is set to 0.005. For the Web data sets, the regularizing parameter C is set to 5 and the power r of the RBF kernel is set to 0.005. For fair comparisons, the parameters of the eSVM are set the same as those of the conventional SVM.

Table 6.2 shows the experimental results of the SVM and the eSVM on the 17 data sets in terms of the number of support vectors (#SV), the classification running time, and the classification rate. Table 6.2 first reveals that the number of support vectors of the eSVM is significantly less than that of the SVM, and consequently the classification speed of the eSVM is much faster than that of the SVM. High computational efficiency is the primary contribution of the eSVM over the SVM. As we discussed in Sect. 6.2, the computational efficiency of the SVM depends on the number of support vectors. Given a classification problem, the larger the number of support vector is, the lower the computational efficiency becomes. The eSVM improves the computational efficiency of the SVM by reducing the number of support vectors. Table 6.2 shows that the number of support vectors of the SVM, as the number of training samples increases, varies in a large range from 785 to 12,165 for the Adult data sets and from 231 to 2,547 for the Web data sets, respectively. This is consistent with our analysis in Sect. 6.2 that the number of support vectors increases dramatically as the problem becomes more complex, as all the training samples on the wrong side of their margin become support vectors due to the introduction of the slack variables for the conventional soft-margin SVM method. However, the number of support vectors of the eSVM, as the number of training samples increases, varies in a smaller range from 63 to 265 for the Adult data sets and from 65 to 289 for the Web data sets, respectively. Consequently, the classification speed of the eSVM is much faster than that of the SVM. Take the A9a from the Adult data sets for an example. The number of support vectors of the SVM is 12,165 and the running time is 156.15 s. In comparison, the number of support vectors of the eSVM is only 265, which is 97% less than that of the SVM, and the running time is only 3.72 s, which is 41 times faster than that of the SVM.

Table 6.2 also reveals that the eSVM has comparable classification performance to — sometimes a little bit lower and sometimes a little bit higher than — that of the

Table 6.2 Performance evaluation of the SVM and the eSVM

Data set		#training samples	#testing samples	#SV		Time (s)		Rate (%)	
				SVM	eSVM	SVM	eSVM	SVM	eSVM
Adult	A1a	1,605	30,956	785	63	17.09	1.90	82.66	82.65
	A2a	2,205	30,296	1,105	72	27.56	2.38	83.46	83.41
	A3a	3,185	29,376	1,451	87	33.99	2.62	83.55	83.53
	A4a	4,781	27,780	2,091	97	51.44	2.91	83.74	83.86
	A5a	6,414	26,147	2,741	111	56.77	2.90	84.06	84.10
	A6a	11,221	21,341	4,480	156	76.36	3.11	84.06	84.10
	A7a	16,101	16,461	6,324	184	85.53	2.78	84.43	84.48
	A8a	22,697	9,865	8,728	225	69.83	2.10	84.89	84.87
	A9a	32,562	16,281	12,165	265	156.15	3.72	84.89	84.81
Web	W1a	2,477	47,272	231	65	6.65	2.69	97.33	97.38
	W2a	3,470	46,279	300	65	8.20	2.72	97.36	97.39
	W3a	4,912	44,837	361	95	9.36	3.38	97.44	97.49
	W4a	7,366	42,383	510	116	11.97	3.60	97.73	97.85
	W5a	9,888	39,861	629	120	14.09	3.54	97.80	97.84
	W6a	17,188	32,561	1,079	136	18.40	3.30	98.15	98.17
	W7a	24,692	25,057	1,444	164	18.67	2.84	98.28	98.34
	W8a	49,749	14,951	2,547	289	19.80	2.83	98.44	98.53

conventional SVM. As we discussed in Sect. 6.2, the eSVM improves the computational efficiency upon the SVM without sacrificing its generalization performance. The eSVM achieves this by simulating the maximal margin separating boundary of the conventional SVM using fewer support vectors. Therefore, the eSVM maintains a similar separating boundary with the SVM, and subsequently has comparable classification performance with the SVM. Table 6.2 shows that the eSVM has a little bit higher classification rate than that of the SVM for 12 out of the 17 data sets (e.g., A4a and W8a), and a little bit lower rate than that of the SVM for the remaining 5 data sets (e.g., A1a and A9a). The difference on the classification rate between the SVM and the eSVM is in the range of -0.08% (for A9a) and $+0.12\%$ (for W4a).

We finally compare our eSVM method with other simplified SVM methods, such as the Reduced SVM (RSVM) [16]. 6 publicly available large-scale data sets [8] are used as done in [16]: *dna*, *satimage*, *letter*, and *shuttle*, *ijcnn1*, and *protein*. The first four data sets are from the Statlog collection, the fifth data set is from the 2001 IJCNN challenge competition, and the last one is from the UCI collection. The feature values of the samples in the data sets are normalized to $[-1, 1]$ as done in [16]. Only the RBF kernel $K(x_i, x_j) = e^{-r\|x_i - x_j\|^2}$ is applied as done in [16]. The regularizing parameter C and the power r of the RBF kernel are set the same as those in [16] for the conventional soft-margin SVM and the RSVM, respectively. For fair comparisons, the parameters of the eSVM are set the same as those of the conventional soft-margin SVM. Table 6.3 shows the number of training samples, the number of testing samples, the number of classes, and the number of features for each data set, as well as the parameter settings for the SVM, the RSVM, and the eSVM, respectively.

Table 6.4 shows the experimental results of the SVM, the RSVM, and the eSVM on the 6 data sets in terms of the number of support vectors (#SV), the classification accuracy (rate), and the running time (T). Note that the results for the RSVM are from the best reported results in [16].

Table 6.4 first reveals that the number of support vectors for the eSVM is much smaller than that for both the SVM and the RSVM on average. Although the eSVM generates a little bit more support vectors than the RSVM for the *dna* and *protein* data

Table 6.3 Data set description and parameter settings

Data set	#training samples	#testing samples	#class	#features	(C, r)		
					SVM	RSVM	eSVM
dna	2,000	1,186	3	180	$2^4, 2^{-6}$	$2^2, 2^{-6}$	$2^4, 2^{-6}$
Satimage	4,435	2,000	6	36	$2^4, 2^0$	$2^3, 2^0$	$2^4, 2^0$
Letter	15,000	5,000	26	16	$2^4, 2^2$	$2^5, 2^1$	$2^4, 2^2$
Shuttle	43,500	14,500	7	9	$2^{11}, 2^3$	$2^{11}, 2^3$	$2^{11}, 2^3$
Ijcnn1	49,990	91,701	2	22	$2^1, 2^1$	$2^0, 2^0$	$2^1, 2^1$
Protein	17,766	6,621	3	357	$2^1, 2^{-3}$	$2^1, 2^{-3}$	$2^1, 2^{-3}$

Table 6.4 Performance assessment of SVM, RSVM, and eSVM (T stands for time in seconds)

Data set	SVM			RSVM			eSVM		
	#SV	Rate	T	#SV	Rate	T	#SV	rate	T
dna	973	95.45	2.39	372	92.33	1.52	503	95.86	1.03
Satimage	1,611	91.3	2.50	1,826	90	11.4	299	91.7	0.58
Letter	8,931	97.78	28.93	13,928	95.9	149.77	522	97.98	1.73
Shuttle	280	99.92	1.65	4,982	99.81	74.82	96	99.95	0.81
Ijcnn1	5,200	96.14	227.68	200	96.77	6.36	82	97.02	4.60
Protein	17,424	68.51	589.58	596	66.24	35	2,866	69.15	99.38

sets, it outperforms the RSVM in the other four data sets. Take the *letter* data set for an example, the eSVM generates 522 support vectors, while the conventional soft-margin SVM and the RSVM generate 8,931 and 13,928 support vectors, respectively, in comparison. On the average, the eSVM reduces the number of support vectors by 87.31 and 80.06%, respectively, when compared with the conventional SVM and the RSVM methods. As a result, our eSVM method reveals higher computational efficiency than both the conventional soft-margin SVM and the RSVM methods. On the average, the eSVM is 7.9 times faster than the SVM. Note that the running time for the RSVM listed in Table 6.4 is from the paper [16], where the RSVM might be implemented on different system environment, hence, we do not make detailed comparisons.

Table 6.4 also reveals that the eSVM achieves better classification accuracy than the RSVM method. Note that four different implementations of the RSVM method are reported in [16] with different classification results, and the best results are selected to show in Table 6.4. The experimental results on the 6 data sets demonstrate that the RSVM method reduces the number of support vectors at the expense of accuracy to some extent. The classification accuracy for the RSVM method, on the average, is 1.34% lower than that for the conventional SVM. Our eSVM, on the other hand, not only significantly reduces the number of support vectors but also achieves comparable classification accuracy with the conventional SVM. In particular, Table 6.4 shows that the average classification rate of the eSVM is 0.43% higher than that of the SVM method and 1.77% higher than that of the RSVM method, respectively.

6.4.2   Performance Evaluation of the Eye Localization Method

We now evaluate the effectiveness and the efficiency of our eye localization method using 12,776 Face Recognition Grand Challenge (FRGC) images from the FRGC version 2 database [22, 27]. Note that the FRGC images possess challenge characteristics, such as large variations in illumination, skin color (white, yellow, and black), facial expression (eyes open, partially open, or closed), as well as scale and orientation. Additional challenges include eye occlusion caused by eye glasses or long hair, and the red eye effect due to the photographic effect. All these challenge factors increase the difficulty of accurate eye-center localization. We also implement the experiments using the FERET database [28] in order to evaluate the robustness of our proposed method and to compare with some recent eye detection methods. The experimental results on the FERET database are shown in the end of this section.

For the training data, we collect from various sources 3,000 pairs of eyes and 12,000 non-eye patches in our experiments. The effect of illumination variations is alleviated by the illumination normalization process introduced in Sect. 6.3. Figure 6.9 shows some training samples used in our experiments.

Fig. 6.9 Eye (the first row) and non-eye (the second row) training samples

We evaluate our eye localization method in terms of recall and precision. Recall, which is also known as the true positive rate, is defined as the number of the true positives divided by the sum of the true positives and false negatives. Precision is defined as the number of the true positives divided by the sum of the true positives and false positives. A robust detection system normally possesses the property of both high recall and precision.

The parameters of our eye localization system are optimized on a validation set by considering both accuracy and efficiency. Specifically, 60 candidates are chosen through the eye candidate selection stage. 2D Haar basis functions for V^5 are used to derive the 2D Haar wavelet features. As $V^5 = V^0 \oplus W^0 \oplus W^1 \oplus W^2 \oplus W^3 \oplus W^4$, the size of the 2D Haar wavelet features is 1,024. 80 eigenvectors out of 1,024 Haar features are derived using the PCA approach. Only the RBF kernel $K(x_i, x_j) = e^{-r\|x_i-x_j\|^2}$ is used. The parameter r is set to 0.0125. The regularizing parameter C is set to 1.

As discussed in Sect. 6.2, the main advantage of the eSVM over the SVM is the computational efficiency. We therefore first evaluate the computational efficiency of the eSVM in comparison with the SVM. Table 6.5 shows the comparison of the computational efficiency between the SVM and the eSVM using the FRGC database. Actually, Table 6.5 reveals that the eSVM significantly reduces the number of support vectors and as a result increases the detection speed. In particular, the number of support vectors of the eSVM is 97.23% less that that of the SVM. As the number of support vectors decreases, the detection time is reduced. The SVM takes 2.98 s (0.33 images per second) on average to process each image. The eSVM, in comparison, significantly improves the computational efficiency to real-time eye detection. Specifically, the eSVM, which takes 0.15 s (6.67 images per second) on average to process each image, is 20 times faster than the SVM.

We then evaluate the classification performance under the difference choice of Q between the SVM and the eSVM classifiers. As discussed in the end of Sect. 6.3.2, we treat the first Q left and right eye candidates with the largest decision values by the SVM (or eSVM) classifier as the detected eyes. Figure 6.10 shows the comparison of eye detection performance of the SVM and the eSVM in terms of recall and precision using the FRGC database, respectively. We evaluate the performance as the Q varies

Table 6.5 Efficiency comparison between SVM and eSVM

Method	#SV	Total time (s)	Time per image (s)
Haar-SVM	9,615	38,072	2.98
Haar-eSVM	267	1,916	0.15

Fig. 6.10 Recall and
precision of the Haar-SVM
and the Haar-eSVM as Q
varies

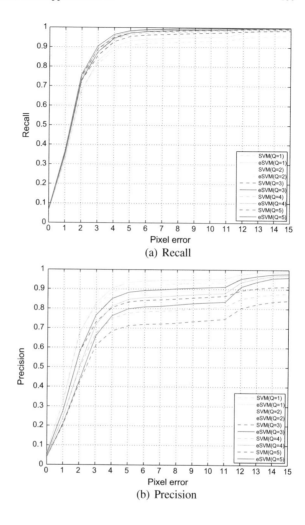

(a) Recall

(b) Precision

from 1 to 5. When $Q > 5$, although recall can further increase, precision decreases
dramatically. The horizontal axis represents the localization pixel errors, and the
vertical axis denotes the accumulated distribution, which means recall (or precision)
of eyes with smaller pixel error than the corresponding horizontal value. Please note
that when $Q = 1$, the recall is equal to the precision according to their definition.

Figure 6.10 shows that the performance of the eSVM tends to be a little bit better
than that of the SVM. In terms of recall, the performance of the eSVM is higher
than SVM on average by 1.28% when $Q = 1$. As Q increases, the difference of the
performance between SVM and eSVM becomes smaller. When $Q = 5$, the performance of the eSVM is only 0.42% higher than that of the SVM on average. In terms
of precision, the difference in performance between SVM and eSVM is more significant. When $Q = 1$, the performance of the eSVM is 1.28% higher than the SVM

on average. The difference increases as Q increases. When $Q = 5$, the performance of the eSVM is 5.25% higher than SVM on average.

Figure 6.10 reveals as well the relationship between the value of Q and the eye detection performance. In particular, Fig. 6.10 shows that as the value of Q becomes larger, recall increases and precision decreases. Actually, if we lower the selection threshold and allow more positive detections, the eye detection rate is increased, but of course the false positives are increased as well. As a matter of fact, if we further increase the value of Q, recall of detections within five pixels of the ground truth can be more than 99%. Precision, however, decreases dramatically to as low as 65%.

We next evaluate the final eye-center localization performance using the FRGC database following the steps introduced at the end of Sect. 6.3.2. In particular, we give only the experimental results of the eSVM based method, which yields better classification performance as shown in Fig. 6.10. Only the recall criterion (i.e., detection rate) is applied, since precision is equal to recall in the case of the single detection for each eye. There are two parameters in the final eye localization: one is the size of the square (i.e., n) and the other is the value of Q, which means how many multiple detections are allowed to choose the final eye center location. Based on the size of the pupil of the normalized training eye sampled, we set $n = 5$. For Q, we search the optimal choice from one to five, since precision will dramatically decrease if Q is greater than five. Figure 6.11a, which shows the final detection rate as Q varies from one to five, indicates that the eye detection performance peaks when $Q = 3$. Table 6.6 shows specific final eye detection rate for each Q value if the eye is considered to be detected correctly when the Euclidean distance between the detected eye center and the ground truth is within five pixels. The eye detection performance when $Q = 3$ is 95.21%.

We therefore use three left and right eye candidates, respectively, to determine the final eye location. Figure 6.11b shows the distribution of the Euclidean distance of detected eyes compared with the ground truth. The average Euclidean distance between the detected eyes and the ground truth is about 2.61 pixels.

In order to assess the robustness of our proposed method and compare with some recent eye detection methods, we finally implement experiments using the FERET database [28]. The FERET datbase contains over 3,300 frontal color face images of nearly 1,000 subjects. The methods we compare with include the HOG descriptor based method by Monzo et al. [24], the hybrid classifier method by Jin et al. [13], the general-to-specific method by Campadelli et al. [3], and a facial identification software—verilook [24, 38]. All the above methods applied the normalized error to evaluate the performance, which is defined as the detection pixel error normalized by the interocular distance. For fair comparisons, we also apply this criterion to evaluate our proposed method. Table 6.7 shows the performance comparison between our eSVM based method and the methods mentioned above for the normalized error of 0.05, 0.10, and 0.25, respectively. Table 6.7 reveals that for the normalized error of 0.05, the detection accuracy of our method is 4.22% higher than the best result reported by the other methods; for the normalized errors of 0.10 and 0.25, the detection accuracy of our method is 2.55 and 1.08% lower than the best results reported by other methods, respectively. Note that the normalized errors of 0.10 and 0.25 are

Fig. 6.11 a Performance comparison of final eye localization under different Q. **b** Distribution of eye localization pixel errors for final eye localization when $Q = 3$

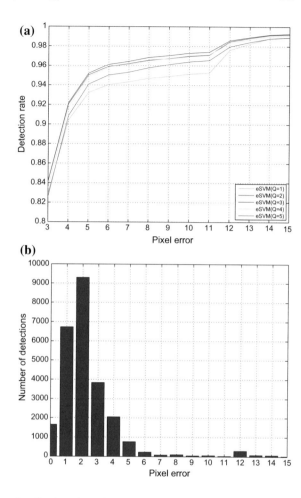

Table 6.6 Performance of final eye localization within five pixel localization error under different Q

Method	Q = 1	Q = 2	Q = 3	Q = 4	Q = 5
Haar-eSVM	93.24	95.03	95.21	94.89	94.09

considered loose criteria which may not be appropriate for evaluating the precise eye detection methods. As a matter of fact, the normalized error of 0.05 is a strict criterion and appropriate for evaluating the precise eye detection methods.

Regarding the efficiency, not many papers report the execution time of their methods. As a matter of fact, speed is an important factor in the real-word application of an eye localization method. Campadelli [3] presented an SVM based eye localization method and reported the execution time of 12 s per image (Java code running on a Pentium 4 with 3.2 GHz). In comparison, the average execution time of our method

Table 6.7 Comparisons of the eye detection performance for different methods on the FERET database. (Note that e stands for the normalized error.)

Method	$e \leq 0.05$ (%)	$e \leq 0.10$ (%)	$e \leq 0.25$ (%)
Monzo [24]	78.00	96.20	99.60
Jin [13]	55.10	93.00	99.80
Campadelli [3]	67.70	89.50	96.40
verilook [38]	74.60	96.80	99.90
Our method	82.22	94.25	98.82

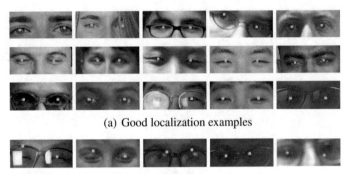

(a) Good localization examples

(b) Bad localization examples

Fig. 6.12 Examples of good (**a**) and bad (**b**) localizations on the FRGC database and the FERET database using our eSVM based method

is only 0.15 s per image due to the application of the eSVM (MATLAB code running on a Pentium 3 with 3.0 GHz). In fact, the execution time can be further significantly reduced if some faster programming languages (like Java or C/C++) and multi-thread techniques are applied.

Figure 6.12 shows some examples of good and bad localizations on both the FRGC and the FERET database using our eSVM based eye localization method.

6.5 Conclusion

We present in this chapter an efficient Support Vector Machine (eSVM) for image search and video retrieval in general and accurate and efficient eye search in particular. Being an efficient and general learning and recognition method, the eSVM can be broadly applied to various tasks in intelligent image search and video retrieval. In particular, we demonstrated in this chapter its applications to accurate and efficient eye localization. Experiments on several diverse data sets show that the eSVM significantly improves the computational efficiency upon the conventional SVM while achieving comparable classification performance with the SVM. Furthermore, the eye localization results on the FRGC and the FERET databases reveal that the proposed eSVM based eye localization method achieves real-time eye detection speed and better eye detection performance than some recent eye detection methods.

References

1. Burges, C.: Simplified support vector decision rule. In: IEEE International Conference on Machine Learning (1996)
2. Burrus, C., Gopinath, R., Guo, H.: Introduction to Wavelets and Wavelet Transforms: A Primer. Prentice-Hall, New Jersey (1998)
3. Campadelli, P., Lanzarotti, R., Lipori, G.: Precise eye localization through a general-to-specific model definition. In: British Machine Vision Conference (2006)
4. Chen, J., Chen, C.: Reducing svm classification time using multiple mirror classifers. IEEE Trans. Syst. Man Cybern. **34**(2), 1173–1183 (2004)
5. Chen, J.H., Chen, C.S.: Reducing svm classification time using multiple mirror classifiers. IEEE Trans. Syst. Man Cybern. Part B **34**(2), 1173–1183 (2004)
6. Chen, P., Lin, C., Scholkopf, B.: A tutorial on υ-support vector machines. Appl. Stoch. Models Bus. Ind. **21**, 111–136 (2005)
7. Chen, S., Liu, C.: Eye detection using color information and a new efficient SVM. In: IEEE International Conferences on Biometrics: Theory, Applications and Systems (2010)
8. Datasets (2016). http://www.csie.ntu.edu.tw/~cjlin/libsvmtools/datasets
9. Davenport, M.A., Baraniuk, R.G., Scott, C.: Tuning support vector machines for minimax and neyman-pearson classification. IEEE Trans. Pattern Anal. Mach. Intell. **32**(10), 1888–1898 (2010)
10. Eckhardt, M., Fasel, I., Movellan, J.: Towards practical facial feature detection. Int. J. Pattern Recognit. Artif. Intell. **23**(3), 379–400 (2009)
11. Haasdonk, B.: Feature space interpretation of svms with indefinite kernels. IEEE Trans. Pattern Anal. Mach. Intell. **27**(4), 482–492 (2005)
12. Heisele, B., Ho, P., Poggio, T.: Face recognition with support vector machines: Global versus component-based approach. In: IEEE International Conference on Computer Vision, pp. 688–694 (2001)
13. Jin, L., Yuan, X., Satoh, S., Li, J., Xia, L.: A hybrid classifier for precise and robust eye detection. In: IEEE International Conference on Pattern Recognition (2006)
14. Kroon, B., Maas, S., Boughorbel, S., Hanjalic, A.: Eye localization in low and standard definition content with application to face matching. Comput. Vis. Image Underst. **113**(4), 921–933 (2009)
15. Lee, Y., Mangasarian, O.: Rsvm: Reduced support vector machines. In: The First SIAM International Conference on Data Mining (2001)
16. Lin, K., Lin, C.: A study on reduced support vector machine. IEEE Trans. Neural Netw. **14**(6), 1449–1559 (2003)
17. Liu, C.: A bayesian discriminating features method for face detection. IEEE Trans. Pattern Anal. Mach. Intell. **25**(6), 725–740 (2003)
18. Liu, C.: Capitalize on dimensionality increasing techniques for improving face recognition grand challenge performance. IEEE Trans. Pattern Anal. Mach. Intell. **28**(5), 725–737 (2006)
19. Liu, C.: The bayes decision rule induced similarity measures. IEEE Trans. Pattern Anal. Mach. Intell. **29**(6), 1086–1090 (2007)
20. Liu, C.: Learning the uncorrelated, independent, and discriminating color spaces for face recognition. IEEE Trans. Inf. Forens. Secur. **3**(2), 213–222 (2008)
21. Liu, C., Wechsler, H.: Probabilistic reasoning models for face recognition. In: Proceedings Computer Vision and Pattern Recognition, pp. 827–832. Santa Barbara, CA (1998)

22. Liu, C., Yang, J.: ICA color space for pattern recognition. IEEE Trans. Neural Netw. **20**(2), 248–257 (2009)
23. Mallat, S.: A theory for multiresolution signal decomposition: the wavelet representation. IEEE Trans. Pattern Anal. Mach. Intell. **11**(7), 674–693 (1989)
24. Monzo, D., Albiol, A., Sastre, J., Albiol, A.: Precise eye localization using hog descriptors. Mach. Vis. Appl. **22**(3), 471–480 (2011)
25. Nguyen, M., Perez, J., Frade, F.: Facial feature detection with optimal pixel reduction svm. In: IEEE International Conference on Automatic Face and Gesture (2008)
26. Osuna, E., Girosi, F.: Reducing the run-time complexity of support vector machines (1998)
27. Phillips, P., Flynn, P., Scruggs, T.: Overview of the face recognition grand challenge. In: IEEE International Conference on Computer Vision and Pattern Recognition (2005)
28. Phillips, P., Moon, H., Rizvi, S., Rauss, P.: The feret evaluation methodology for face recognition algorithms. IEEE Trans. Pattern Anal. Mach. Intell. **22**(10), 1090–1104 (2000)
29. Platt, J.: Fast training of support vector machines using sequential minimal optimization. In: Advances in Kernel Methods - Support Vector Learning. MIT Press (1998)
30. Platt, J.C.: Sequential Minimal Optimization: A Fast Algorithm for Training Support Vector Machines. MIT Press, Cambridge (1998)
31. Rätsch, G., Mika, S., Schölkopf, B., Müller, K.R.: Constructing boosting algorithms from svms: an application to one-class classification. IEEE Trans. Pattern Anal. Mach. Intell. **24**(9), 1184–1199 (2002)
32. U.M.L. Repository (2016). http://www.ics.uci.edu/~mlearn/MLRepository.html
33. Romdhani, S., Torr, B., Scholkopf, B., Blake, A.: Computationally efficient face detection. In: IEEE International Conference on Computer Vision (2001)
34. Scholkopf, B.E.A.: Input space versus feature space in kernel-based methods. IEEE Trans. Neural Netw. **10**(5), 1000–1017 (1999)
35. Shih, P., Liu, C.: Comparative assessment of content-based face image retrieval in different color spaces. IJPRAI **19**(7), 873–893 (2005)
36. Tan, X., Triggs, B.: Enhanced local texture feature sets for face recognition under difficult lighting conditions. IEEE Trans. Image Process. **19**(6), 1635–1650 (2010)
37. Vapnik, V.N.: The nature of statistical learning theory (2000)
38. Verilook (2016). http://www.neurotechnology.com/verilook.html
39. Wang, P., Ji, Q.: Multi-view face and eye detection using discriminant features. Comput. Vis. Image Underst. **105**(2), 99–111 (2007)

Chapter 7
SIFT Features in Multiple Color Spaces for Improved Image Classification

Abhishek Verma and Chengjun Liu

Abstract This chapter first discusses oRGB-SIFT descriptor, and then integrates it with other color SIFT features to produce the Color SIFT Fusion (CSF), the Color Grayscale SIFT Fusion (CGSF), and the CGSF+PHOG descriptors for image classification with special applications to image search and video retrieval. Classification is implemented using the EFM-NN classifier, which combines the Enhanced Fisher Model (EFM) and the Nearest Neighbor (NN) decision rule. The effectiveness of the proposed descriptors and classification method is evaluated using two large scale and challenging datasets: the Caltech 256 database and the UPOL Iris database. The experimental results show that (i) the proposed oRGB-SIFT descriptor improves recognition performance upon other color SIFT descriptors; and (ii) the CSF, the CGSF, and the CGSF+PHOG descriptors perform better than the other color SIFT descriptors.

7.1 Introduction

Content-based image retrieval is based on image similarity in terms of visual content such as features from color, texture, shape, etc. to a user-supplied query image or user-specified image features has been a focus of interest for the past several years. Color features provide powerful information for image search, indexing, and classification [26, 32, 41], in particular for identification of biometric images [36, 38], objects, natural scene, image texture and flower categories [2, 37, 39] and geographical features from images. The choice of a color space is important for many computer vision algorithms. Different color spaces display different color properties. With the large variety of available color spaces, the inevitable question that arises is how to select a color space that produces best results for a particular computer vision

A. Verma (✉)
California State University, Fullerton, CA 92834, USA
e-mail: averma@fullerton.edu

C. Liu
New Jersey Institute of Technology, Newark, NJ 07102, USA
e-mail: chengjun.liu@njit.edu

© Springer International Publishing AG 2017
C. Liu (ed.), *Recent Advances in Intelligent Image Search and Video Retrieval*,
Intelligent Systems Reference Library 121, DOI 10.1007/978-3-319-52081-0_7

task. Two important criteria for color feature detectors are that they should be stable under varying viewing conditions, such as changes in illumination, shading, highlights, and they should have high discriminative power. Color features such as the color histogram, color texture and local invariant features provide varying degrees of success against image variations such as viewpoint and lighting changes, clutter and occlusions [7, 9, 34].

In the past, there has been much emphasis on the detection and recognition of locally affine invariant regions [27, 29]. Successful methods are based on representing a salient region of an image by way of an elliptical affine region, which describes local orientation and scale. After normalizing the local region to its canonical form, image descriptors are able to capture the invariant region appearance. Interest point detection methods and region descriptors can robustly detect regions, which are invariant to translation, rotation and scaling [27, 29]. Affine region detectors when combined with the intensity Scale-Invariant Feature Transform (SIFT) descriptor [27] has been shown to outperform many alternatives [29].

In this chapter, the SIFT descriptor is extended to different color spaces, including oRGB color space [6], oRGB-SIFT feature representation is proposed, furthermore it is integrated with other color SIFT features to produce the Color SIFT Fusion (CSF), and the Color Grayscale SIFT Fusion (CGSF) descriptors. Additionally, the CGSF is combined with the Pyramid of Histograms of Orientation Gradients (PHOG) to obtain the CGSF+PHOG descriptor for image category classification with special applications to biometrics. Classification is implemented using EFM-NN classifier [24, 25], which combines the Enhanced Fisher Model (EFM) and the Nearest Neighbor (NN) decision rule [12]. The effectiveness of the proposed descriptors and classification method is evaluated on two large scale, grand challenge datasets: the Caltech 256 dataset and the UPOL Iris database.

Rest of the chapter is organized as follows. In Sect. 7.2 we review image-level global and local feature descriptors. Section 7.3 presents a review of five color spaces in which the color SIFT descriptors are defined followed by a discussion on clustering, visual vocabulary tree, and visual words for SIFT descriptors in Sect. 7.4. Thereafter, in Sect. 7.5 five conventional SIFT descriptors are presented: the RGB-SIFT, the rgb-SIFT, the HSV-SIFT, the YCbCr-SIFT, and the grayscale-SIFT descriptors and four new color SIFT descriptors are presented: the oRGB-SIFT, the Color SIFT Fusion (CSF), the Color Grayscale SIFT Fusion (CGSF), and the CGSF+PHOG descriptors. Section 7.6 presents a detailed discussion on the EFM-NN classification methodology. Description of datasets used for evaluation of methodology is provided in Sect. 7.7. Next, in Sect. 7.8 we present experimental results of evaluation of color SIFT descriptors. Section 7.9 concludes the chapter.

7.2   Related Work

In past years, use of color as a means to biometric image retrieval [22, 26, 32] and object and scene search has gained popularity. Color features can capture discriminative information by means of the color invariants, color histogram, color texture, etc. The earliest methods for object and scene classification were mainly based on the global descriptors such as the color and texture histogram [30, 31]. One of the earlier works is the color indexing system designed by Swain and Ballard, which uses the color histogram for image inquiry from a large image database [35]. Such methods are sensitive to viewpoint and lighting changes, clutter and occlusions. For this reason, global methods were gradually replaced by the part-based methods, which became one of the popular techniques in the object recognition community. Part-based models combine appearance descriptors from local features along with their spatial relationship. Harris interest point detector was used for local feature extraction; such features are only invariant to translation [1, 40]. Afterwards, local features with greater invariance were developed, which were found to be robust against scale changes [11] and affine deformations [20]. Learning and inference for spatial relations poses a challenging problem in terms of its complexity and computational cost. Whereas, the orderless bag-of-words methods [11, 17, 21] are simpler and computationally efficient, though they are not able to represent the geometric structure of the object or to distinguish between foreground and background features. For these reasons, the bag-of-words methods are not robust to clutter. One way to overcome this drawback is to design kernels that can yield high discriminative power in presence of noise and clutter [15].

Further, work on color based image classification appears in [23, 26, 41] that propose several new color spaces and methods for face classification and in [5] the HSV color space is used for the scene category recognition. Evaluation of local color invariant descriptors is performed in [7]. Fusion of color models, color region detection and color edge detection have been investigated for representation of color images [34]. Key contributions in color, texture, and shape abstraction have been discussed in Datta et al. [9].

As discussed before, many recent techniques for the description of images have considered local features. The most successful local image descriptor so far is Lowe's SIFT descriptor [27]. The SIFT descriptor encodes the distribution of Gaussian gradients within an image region. It is a 128-bin histogram that summarizes the local oriented gradients over 8 orientations and over 16 locations. This can efficiently represent the spatial intensity pattern, while being robust to small deformations and localization errors. Several modifications to the SIFT features have been proposed; among them are the PCA-SIFT [18], GLOH [28], and SURF [3]. These region-based descriptors have achieved a high degree of invariance to the overall illumination conditions for planar surfaces. Although, designed to retrieve identical object patches, SIFT-like features turn out to be quite successful in the bag-of-words approaches for general scene and object classification [5].

The Pyramid of Histograms of Orientation Gradients (PHOG) descriptor [4] is able to represent an image by its local shape and the spatial layout of the shape. The local shape is captured by the distribution over edge orientations within a region, and the spatial layout by tiling the image into regions at multiple resolutions. The distance between two PHOG image descriptors then reflects the extent to which the images contain similar shapes and correspond in their spatial layout.

7.3   Color Spaces

This section presents a review of five color spaces in which the color SIFT descriptors are defined.

7.3.1   RGB and rgb Color Spaces

A color image contains three component images, and each pixel of a color image is specified in a color space, which serves as a color coordinate system. The commonly used color space is the RGB color space. Other color spaces are usually calculated from the RGB color space by means of either linear or nonlinear transformations.

To reduce the sensitivity of the RGB images to luminance, surface orientation, and other photographic conditions, the rgb color space is defined by normalizing the R, G, and B components:

$$\begin{aligned} r &= R/(R+G+B) \\ g &= G/(R+G+B) \\ b &= B/(R+G+B) \end{aligned} \tag{7.1}$$

Due to the normalization r and g are scale-invariant and thereby invariant to light intensity changes, shadows and shading [13].

7.3.2   HSV Color Space

The HSV color space is motivated by human vision system because humans describe color by means of hue, saturation, and brightness. Hue and saturation define chrominance, while intensity or value specifies luminance [14]. The HSV color space is defined as follows [33]:

$$\begin{aligned} Let \quad & \begin{cases} MAX = max(R, G, B) \\ MIN = min(R, G, B) \\ \delta = MAX - MIN \end{cases} \\ V &= MAX \\ S &= \begin{cases} \frac{\delta}{MAX} & \text{if } MAX \neq 0 \\ 0 & \text{if } MAX = 0 \end{cases} \\ H &= \begin{cases} 60(\frac{G-B}{\delta}) & \text{if } MAX = R \\ 60(\frac{B-R}{\delta} + 2) & \text{if } MAX = G \\ 60(\frac{R-G}{\delta} + 4) & \text{if } MAX = B \\ not\ defined & \text{if } MAX = 0 \end{cases} \end{aligned} \tag{7.2}$$

7.3.3 YCbCr Color Space

The YCbCr color space is developed for digital video standard and television transmissions. In YCbCr, the RGB components are separated into luminance, chrominance blue, and chrominance red:

$$
\begin{bmatrix} Y \\ Cb \\ Cr \end{bmatrix} = \begin{bmatrix} 16 \\ 128 \\ 128 \end{bmatrix} + \begin{bmatrix} 65.4810 & 128.5530 & 24.9660 \\ -37.7745 & -74.1592 & 111.9337 \\ 111.9581 & -93.7509 & -18.2072 \end{bmatrix} \begin{bmatrix} R \\ G \\ B \end{bmatrix} \qquad (7.3)
$$

where the R, G, B values are scaled to $[0, 1]$.

7.3.4 oRGB Color Space

The oRGB color space [6] has three channels L, $C1$ and $C2$. The primaries of this model are based on the three fundamental psychological opponent axes: white-black, red-green, and yellow-blue. The color information is contained in $C1$ and $C2$. The value of $C1$ lies within $[-1, 1]$ and the value of $C2$ lies within $[-0.8660, 0.8660]$. The L channel contains the luminance information and its values range between $[0, 1]$:

$$
\begin{bmatrix} L \\ C1 \\ C2 \end{bmatrix} = \begin{bmatrix} 0.2990 & 0.5870 & 0.1140 \\ 0.5000 & 0.5000 & -1.0000 \\ 0.8660 & -0.8660 & 0.0000 \end{bmatrix} \begin{bmatrix} R \\ G \\ B \end{bmatrix} \qquad (7.4)
$$

7.4 SIFT Feature Extraction, Clustering, Visual Vocabulary Tree, and Visual Words

This section first gives details of the SIFT feature extraction procedure. The next phase deals with the formation of visual vocabulary tree and visual words, here the normalized SIFT features are quantized with the vocabulary tree such that each image is represented as a collection of visual words, provided from a visual vocabulary. The visual vocabulary is obtained by vector quantization of descriptors computed from the training images using k-means clustering. See Fig. 7.1 for an overview of the processing pipeline.

7.4.1 SIFT Feature Extraction

Image similarity may be defined in many ways based on the need of the application. It could be based on shape, texture, resolution, color or some other spatial features.

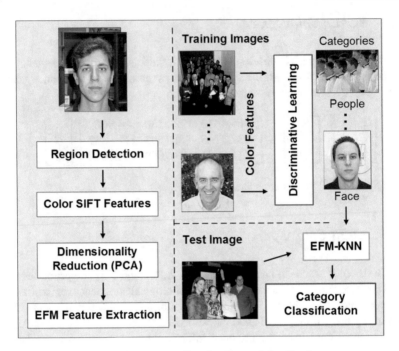

Fig. 7.1 An overview of SIFT feature extraction, learning and classification stages

The experiments here compute the SIFT descriptors extracted from the scale invariant points [42] on aforementioned color spaces. Such descriptors are called sparse descriptors, they have been previously used in [8, 19]. Scale invariant points are obtained with the Hessian-affine point detector on the intensity channel. For the experiments, the Hessian-affine point detector is used because it has shown good performance in category recognition [29]. The remaining portion of feature extraction is then implemented according to the SIFT feature extraction pipeline of Lowe [27]. Canonical directions are found based on an orientation histogram formed on the image gradients. SIFT descriptors are then extracted relative to the canonical directions.

7.4.2 Clustering, Visual Vocabulary Tree, and Visual Words

The visual vocabulary tree defines a hierarchical quantization that is constructed with the hierarchical k-means clustering. A large set of representative descriptor vectors taken from the training images are used in the unsupervised training of the tree. Instead of k defining the final number of clusters or quantization cells, k defines the branch factor (number of children of each node) of the tree. First, an initial k-means process is run on the training data, defining k cluster centers. The training data is then partitioned into k groups, where each group consists of the descriptor vectors closest to a particular cluster center. The same process is then recursively applied to each group of descriptor vectors, recursively defining clusters by splitting each cluster into k new parts. The tree is determined level by level, up to some maximum number of levels say L, and each division into k parts is only defined by the distribution of the descriptor vectors that belong to the parent cluster. Once the tree is computed, its leaf nodes are used for quantization of descriptors from the training and test images.

It has been experimentally observed that most important for the retrieval quality is to have a large vocabulary, i.e., large number of leaf nodes. While the computational cost of increasing the size of the vocabulary in a non-hierarchical manner would be very high, the computational cost in the hierarchical approach is logarithmic in the number of leaf nodes. The memory usage is linear in the number of leaf nodes kL. The current implementation builds a tree of 6, 561 leaf nodes and $k = 9$. See Fig. 7.2 for an overview of the quantization process.

To obtain fixed-length feature vectors per image, the visual words model is used [5, 8]. The visual words model performs vector quantization of the color descriptors in an image against a visual vocabulary. In the quantization phase, each descriptor vector is simply propagated down the tree at each level by comparing the descriptor vector to the k candidate cluster centers (represented by k children in the tree) and choosing the closest one till it is assigned to a particular leaf node. This is a simple matter of performing k dot products at each level, resulting in a total of kL dot products, which is very efficient if k is not too large.

(a)   **(b)**

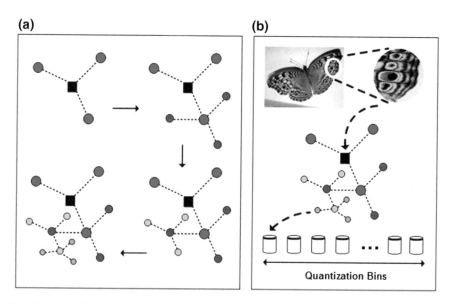

Fig. 7.2 **a** An illustration of the process of constructing a vocabulary tree by hierarchical k-means. The hierarchical quantization is defined at each level by k centers (in this case $k = 3$). **b** A large number of *elliptical* regions are extracted from the image and normalized to *circular* regions. A SIFT descriptor vector is computed for each region. The descriptor vector is then hierarchically quantized by the vocabulary tree. The number of quantization bins is the number of leaf nodes in the vocabulary tree; this is the length of the final feature vector as well

Once all the SIFT features from an image are quantized, a fixed length feature vector would be obtained. The feature vector is normalized to zero mean and unit standard deviation. The advantage of representing an image as a fixed length feature vector lies in the fact that it allows to effectively compare images that vary in size.

7.5 Color SIFT Descriptors

The SIFT descriptor proposed by Lowe transforms an image into a large collection of feature vectors, each of which is invariant to image translation, scaling, and rotation, partially invariant to the illumination changes, and robust to local geometric distortion [27]. The key locations used to specify the SIFT descriptor are defined as maxima and minima of the result of the difference of Gaussian function applied in the scale-space to a series of smoothed and resampled images. SIFT descriptors robust to local affine distortions are then obtained by considering pixels around a radius of the key location.

The grayscale SIFT descriptor is defined as the SIFT descriptor applied to the grayscale image. A color SIFT descriptor in a given color space is derived by individually computing the SIFT descriptor on each of the three component images in the specific color space. This produces a 384 dimensional descriptor that is formed from concatenating the 128 dimensional vectors from the three channels. As a result, four conventional color SIFT descriptors are defined: the RGB-SIFT, the YCbCr-SIFT, the HSV-SIFT, and the rgb-SIFT descriptors.

Furthermore, four new color SIFT descriptors are defined in the oRGB color space and the fusion in different color spaces. In particular, the oRGB-SIFT descriptor is constructed by concatenating the SIFT descriptors of the three component images in the oRGB color space. The Color SIFT Fusion (CSF) descriptor is formed by fusing the RGB-SIFT, the YCbCr-SIFT, the HSV-SIFT, the oRGB-SIFT, and the rgb-SIFT descriptors. The Color Grayscale SIFT Fusion (CGSF) descriptor is obtained by fusing further the CSF descriptor and the grayscale-SIFT descriptor. The CGSF is combined with the Pyramid of Histograms of Orientation Gradients (PHOG) descriptor to obtain the CGSF+PHOG descriptor. See Fig. 7.3 for multiple Color SIFT features fusion methodology.

7.6 EFM-NN Classifier

Image classification using the descriptors introduced in the preceding section is implemented using EFM-NN classifier [24, 25], which combines the Enhanced Fisher Model (EFM) and Nearest Neighbor (NN) decision rule [12]. Let $\mathscr{X} \in \mathbb{R}^N$ be a random vector whose covariance matrix is $\Sigma_{\mathscr{X}}$:

$$\Sigma_{\mathscr{X}} = \mathscr{E}\{[\mathscr{X} - \mathscr{E}(\mathscr{X})][\mathscr{X} - \mathscr{E}(\mathscr{X})]^t\} \tag{7.5}$$

Fig. 7.3 Multiple Color SIFT features fusion methodology using the EFM feature extraction

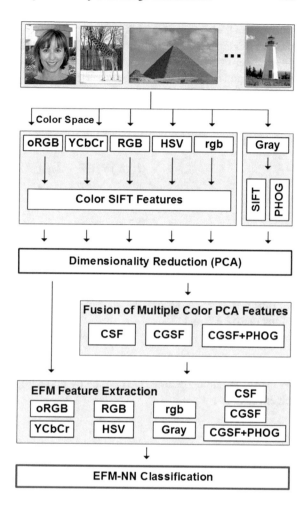

where $\mathscr{E}(\cdot)$ is the expectation operator and t denotes the transpose operation. The eigenvectors of the covariance matrix $\Sigma_{\mathscr{X}}$ can be derived by PCA:

$$\Sigma_{\mathscr{X}} = \Phi \Lambda \Phi^t \qquad (7.6)$$

where $\Phi = [\phi_1 \phi_2 \dots \phi_N]$ is an orthogonal eigenvector matrix and $\Lambda = diag\{\lambda_1, \lambda_2, \dots, \lambda_N\}$ a diagonal eigenvalue matrix with diagonal elements in decreasing order. An important application of PCA is dimensionality reduction:

$$\mathscr{Y} = P^t \mathscr{X} \qquad (7.7)$$

where $P = [\phi_1 \phi_2 \ldots \phi_K]$, and $K < N$. $\mathcal{Y} \in \mathbb{R}^K$ thus is composed of the most significant principal components. PCA, which is derived based on an optimal representation criterion, usually does not lead to good image classification performance. To improve upon PCA, the Fisher Linear Discriminant (FLD) analysis [12] is introduced to extract the most discriminating features.

The FLD method optimizes a criterion defined on the within-class and between-class scatter matrices, S_w and S_b [12]:

$$S_w = \sum_{i=1}^{L} P(\omega_i) \mathcal{E}\{(\mathcal{Y} - M_i)(\mathcal{Y} - M_i)^t | \omega_i\} \tag{7.8}$$

$$S_b = \sum_{i=1}^{L} P(\omega_i)(M_i - M)(M_i - M)^t \tag{7.9}$$

where $P(\omega_i)$ is *a priori* probability, ω_i represent the classes, and M_i and M are the means of the classes and the grand mean, respectively. The criterion the FLD method optimizes is $J_1 = tr(S_w^{-1} S_b)$, which is maximized when Ψ contains the eigenvectors of the matrix $S_w^{-1} S_b$ [12]:

$$S_w^{-1} S_b \Psi = \Psi \Delta \tag{7.10}$$

where Ψ, Δ are the eigenvector and eigenvalue matrices of $S_w^{-1} S_b$, respectively. The FLD discriminating features are defined by projecting the pattern vector $\mathcal{Y}$ onto the eigenvectors of Ψ:

$$\mathcal{Z} = \Psi^t \mathcal{Y} \tag{7.11}$$

$\mathcal{Z}$ thus is more effective than the feature vector $\mathcal{Y}$ derived by PCA for image classification.

The FLD method, however, often leads to overfitting when implemented in an inappropriate PCA space. To improve the generalization performance of the FLD method, a proper balance between two criteria should be maintained: the energy criterion for adequate image representation and the magnitude criterion for eliminating the small-valued trailing eigenvalues of the within-class scatter matrix [24]. Enhanced Fisher Model (EFM), is capable of improving the generalization performance of the FLD method [24]. Specifically, the EFM method improves the generalization capability of the FLD method by decomposing the FLD procedure into a simultaneous diagonalization of the within-class and between-class scatter matrices [24]. The simultaneous diagonalization is stepwise equivalent to two operations as pointed out by [12]: whitening the within-class scatter matrix and applying PCA to the between-class scatter matrix using the transformed data. The stepwise operation shows that during whitening the eigenvalues of the within-class scatter matrix appear in the denominator. Since the small (trailing) eigenvalues tend to capture noise [24], they cause the whitening step to fit for misleading variations, which leads to poor generalization performance. To achieve enhanced performance, the EFM method preserves a proper balance between the need that the selected eigenvalues account for most of the spectral energy of the raw data (for representational adequacy), and the requirement that the eigenvalues of the within-class scatter matrix (in the reduced PCA space) are not too small (for better generalization performance) [24].

Image classification is implemented with EFM-NN, which uses nearest neighbor and cosine distance measure. Figure 7.3 shows the fusion methodology of multiple descriptors using EFM feature extraction and EFM-NN classification.

7.7   Description of Dataset

We perform experimental evaluation of Color SIFT descriptors on two publicly available large scale grand challenge datasets: the Caltech 256 object categories dataset and the UPOL iris dataset.

7.7.1   Caltech 256 Object Categories Dataset

The Caltech 256 dataset [16] comprises of 30,607 images divided into 256 categories and a clutter class. See Fig. 7.4 for some images from the object categories and Fig. 7.5 for some sample images from the Faces and People categories. The images have high intra-class variability and high object location variability. Each category contains at least 80 images, a maximum of 827 images and the mean number of images per

Fig. 7.4 Example images from the Caltech 256 object categories dataset

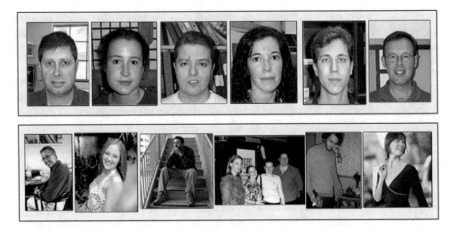

Fig. 7.5 Example images from the faces and people classes of the Caltech 256 object categories dataset

category is 119. The images have been collected from Google and PicSearch, they represent a diverse set of lighting conditions, poses, back-grounds, image sizes, and camera systematics. The various categories represent a wide variety of natural and artificial objects in various settings. The images are in color, in JPEG format with only a small number of grayscale images. The average size of each image is 351×351 pixels.

7.7.2 UPOL Iris Dataset

The UPOL iris dataset [10] contains 128 unique eyes (or classes) belonging to 64 subjects with each class containing three sample images. The images of the left and right eyes of a person belong to different classes. The irises were scanned by a TOPCON TRC50IA optical device connected with a SONY DXC-950P 3CCD camera. The iris images are in 24-bit PNG format (color) and the size of each image is 576×768 pixels. See Fig. 7.6 for some sample images from this dataset.

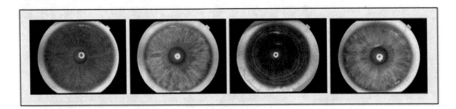

Fig. 7.6 Example images from the UPOL Iris dataset

7.8   Experimental Evaluation of Color SIFT Descriptors on the Caltech 256 and the UPOL Iris Datasets

7.8.1   Experimental Methodology

In order to make a comparative assessment of the descriptors and methods; from the aforementioned two datasets we the Biometric 100 dataset with 100 categories includes the Iris category from the UPOL dataset, Faces and People categories and 97 randomly chosen categories from the Caltech 256 dataset. This dataset is of high difficulty due to the large number of classes with high intra-class and low inter-class variations.

The classification task is to assign each test image to one of a number of categories. The performance is measured using a confusion matrix, and the overall performance rates are measured by the average value of the diagonal entries of the confusion matrix. Dataset is split randomly into two separate sets of images for training and testing. From each class 60 images for training and 20 images for testing are randomly selected. There is no overlap in the images selected for training and testing. The classification scheme on the dataset compares the overall and category wise performance of ten different descriptors: the oRGB-SIFT, the YCbCr-SIFT, the RGB-SIFT, the HSV-SIFT, the rgb-SIFT, the PHOG, the grayscale-SIFT, the CSF, the CGSF, and the CGSF+PHOG descriptors. Classification is implemented using EFM-NN classifier, which combines the Enhanced Fisher Model (EFM) and the Nearest Neighbor (NN) decision rule.

7.8.2   Experimental Results on the Biometric 100 Categories Dataset

7.8.2.1   Evaluation of Overall Classification Performance of Descriptors with the EFM-NN Classifier

The first set of experiments assesses the overall classification performance of the ten descriptors on the Biometric 100 dataset with 100 categories. Note that for each category a five-fold cross validation is implemented for each descriptor using the EFM-NN classification technique to derive the average classification performance. As a result, each descriptor yields 100 average classification rates corresponding to the 100 image categories. The mean value of these 100 average classification rates is defined as the mean average classification performance for the descriptor.

The best recognition rate that is obtained is 51.9% from the CGSF+PHOG, which is a very respectable value for a dataset of this size and complexity. The oRGB-SIFT achieves the classification rate of 32.2% and hence once again outperforms

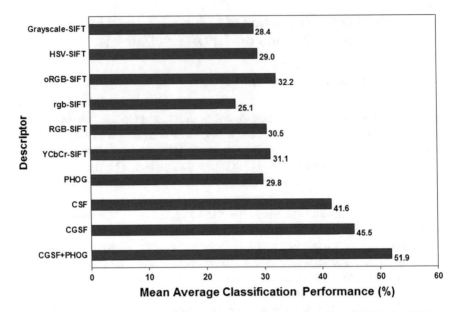

Fig. 7.7 The mean average classification performance of the ten descriptors: the oRGB-SIFT, the YCbCr-SIFT, the RGB-SIFT, the HSV-SIFT, the rgb-SIFT, the grayscale-SIFT, the PHOG, the CSF, the CGSF, and the CGSF+PHOG descriptors on the Biometric 100 dataset

other color descriptors. The success rate for YCbCr-SIFT comes in second place with 31.1% followed by the RGB-SIFT at 30.5%. Fusion of color SIFT descriptors (CSF) improves over the grayscale-SIFT by a huge 13.2%. Again, the grayscale-SIFT shows more distinctiveness than the rgb-SIFT, and improves the fusion (CGSF) result by a good 3.9% over the CSF. Fusing the CGSF and PHOG further improves the recognition rate over the CGSF by 6.4%. See Fig. 7.7 for mean average classification performance of various descriptors.

7.8.2.2 Comparison of PCA and EFM-NN Results

The second set experiments compares the classification performance of the PCA and the EFM-NN (nearest neighbor) classifiers. Table 7.1 shows the results of the two classifiers across various descriptors. It can be seen that the EFM-NN technique improves over the PCA technique by 2–3% on the color SIFT descriptors, by 2.1% on the grayscale-SIFT, and by 1.9% on the PHOG. The improvement on fused descriptors is in the range of 1–2.6%. These results reaffirm the superiority of the EFM-NN classifier over the PCA technique.

Table 7.1 Comparison of classifiers across ten descriptors (%) on the biometric 100 dataset

Descriptor	PCA	EFM-NN
RGB-SIFT	27.9	**30.5**
HSV-SIFT	26.1	**29.0**
rgb-SIFT	23.1	**25.1**
oRGB-SIFT	29.4	**32.2**
YCbCr-SIFT	28.2	**31.1**
SIFT	26.3	**28.4**
PHOG	28.0	**29.8**
CSF	40.2	**41.6**
CGSF	44.6	**45.5**
CGSF+PHOG	49.4	**51.9**

7.8.2.3 Evaluation of PCA and EFM-NN Results upon Varying Number of Features

The third set of experiments evaluates the classification performance using the PCA and the EFM-NN methods respectively by varying the number of features over the following ten descriptors: CGSF+PHOG, CGSF, CSF, YCbCr-SIFT, oRGB-SIFT, RGB-SIFT, HSV-SIFT, Grayscale-SIFT, rgb-SIFT, and PHOG.

Classification performance is computed for up to 780 features with the PCA classifier. From Fig. 7.8 it can be seen that the success rate for the CGSF+PHOG stays consistently above that of the CGSF and CSF over varying number of features and peaks at around 660 features. These three descriptors show an increasing trend overall and flatten out toward the end. The oRGB-SIFT, YCbCr-SIFT, RGB-SIFT, and grayscale-SIFT show a similar increasing trend and flatten toward the end. The oRGB-SIFT descriptor consistently stays above other color SIFT descriptors. The HSV-SIFT and PHOG peak in the first half of the graph and show a declining trend thereafter. The grayscale-SIFT maintains its superior performance upon the rgb-SIFT on the varying number of features.

With the EFM-NN classifier, the success rates are computed for up to 95 features. From Fig. 7.9 it can be seen that the success rate for the CGSF+PHOG stays consistently above that of the CGSF and CSF over varying number of features and peaks at about 80 features. These three descriptors show an increasing trend throughout and tend to flatten above 65 features. The oRGB-SIFT consistently stays above the rest of the descriptors. The grayscale-SIFT improves over the rgb-SIFT but falls below the PHOG.

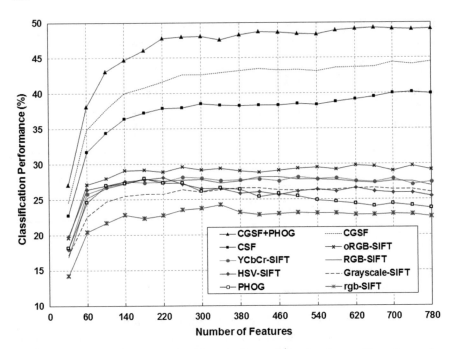

Fig. 7.8 Classification results using the PCA method across the ten descriptors with varying number of features on the Biometric 100 dataset

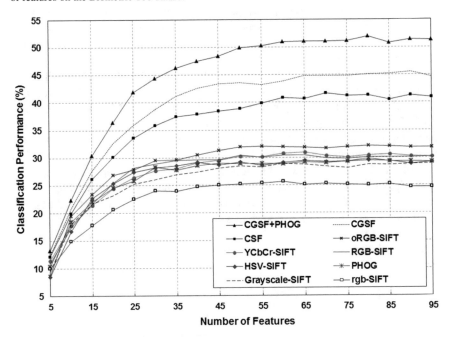

Fig. 7.9 Classification results using the EFM-NN method across the ten descriptors with varying number of features on the Biometric 100 dataset

7.8.2.4   Evaluation of Descriptors and Classifier on Individual Image Categories

The fourth set of experiments assesses the eight descriptors using the EFM-NN classifier on individual image categories. Here a detailed analysis of the performance of the descriptors is performed with the EFM-NN classifier over 100 image categories. First the classification results on the three biometric categories are presented. From Table 7.2 it can be seen that the Iris category has a 100% recognition rate across all the descriptors. For the Faces category the color SIFT descriptors outperform the grayscale-SIFT by 5–10% and the fusion of all descriptors (CGSF+PHOG) reaches a 95% success rate. The People category achieves a high success rate of 40% with the CGSF+PHOG, surprisingly grayscale-SIFT outperforms the color descriptors by 10–20%. The fusion of individual SIFT descriptors (CGSF) improves the classification performance for the People category.

Table 7.2   Category wise descriptor performance (%) split-out with the EFM-NN classifier on the biometric 100 dataset (Note that the categories are sorted on the CGSF+PHOG results)

Category	CGSF+PHOG	CGSF	CSF	oRGB SIFT	YCbCr SIFT	RGB SIFT	Gray SIFT	PHOG
Iris	**100**	**100**	**100**	**100**	**100**	**100**	**100**	**100**
Faces	**95**	90	90	90	**95**	90	85	**95**
People	**40**	**40**	25	20	20	15	30	10
Hibiscus	**100**	**100**	95	70	80	85	75	55
French horn	**95**	85	85	85	65	80	90	20
Leopards	95	90	**100**	90	95	95	**100**	90
Saturn	**95**	**95**	**95**	**95**	85	90	**95**	55
School bus	**95**	**95**	**95**	75	85	**95**	80	60
Swiss army knife	**95**	90	80	65	75	65	65	25
Watch	**95**	60	55	45	40	45	30	85
Zebra	**95**	80	60	60	35	40	45	60
Galaxy	**90**	85	85	85	70	65	80	15
American flag	**85**	**85**	80	55	75	65	40	5
Cartman	**85**	75	75	40	55	65	55	30
Desk-globe	**85**	75	75	60	65	65	45	80
Harpsichord	**85**	80	**85**	50	80	70	60	55
Ketch	**85**	**85**	**85**	45	50	45	50	70
Roulette wheel	**85**	80	75	70	65	75	55	35
Hawksbill	**80**	**80**	75	55	60	70	55	40
Iris flower	**80**	75	75	35	65	**80**	65	30
Mountain bike	80	85	**90**	70	65	85	75	70

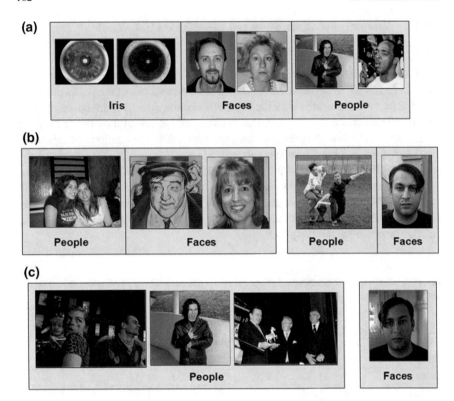

Fig. 7.10 Image recognition using the EFM-NN classifier on the Biometric 100 dataset: **a** examples of the correctly classified images from the three biometric image categories; **b** images unrecognized using the grayscale-SIFT descriptor but recognized using the oRGB-SIFT descriptor; **c** images unrecognized using the oRGB-SIFT descriptor but recognized using the CSF descriptor; **d** images unrecognized using the CSF but recognized using the CGSF+PHOG; **e** images unrecognized by PCA but recognized by EFM-NN on the CGSF+PHOG descriptor

The average success rate for the CGSF+PHOG over the top 20 categories is 90% with ten categories above the 90% mark. Individual color SIFT features improve upon the grayscale-SIFT on most of the categories, in particular for the Swiss army knife, Watch, American flag, and Roulette wheel categories. The CSF almost always improves over the grayscale-SIFT, with the exception of People and French horn categories. The CGSF either is at par or improves over the CSF for all categories with the exception of two of the categories. Most categories perform at their best when the PHOG is combined with the CGSF.

7.8.2.5   Evaluation of Descriptors and Classifier Based on Correctly Recognized Images

The final set of experiments further assesses the performance of the descriptors based on the correctly recognized images. See Fig. 7.10a for some examples of the correctly classified images from the Iris, Faces, and People categories. Once again notice the high intra-class variability in the recognized images for the Faces and People class. Figure 7.10b shows some images from the Faces and People categories that are not recognized by the grayscale-SIFT but are correctly recognized by the oRGB-SIFT. Figure 7.10c shows some images that are not recognized by the oRGB-SIFT but are correctly recognized by the CSF. Figure 7.10d shows some images from the People class, which are not recognized by the CSF but are correctly recognized by the CGSF+PHOG descriptor. Thus, combining grayscale-SIFT, PHOG, and CSF lends more discriminative power. Lastly in Fig. 7.10e a face image unrecognized by the PCA but recognized by the EFM-NN classifier on the CGSF+PHOG descriptor.

See Fig. 7.11a for some examples of the images unrecognized by the gray-scale-SIFT but are correctly recognized by the oRGB-SIFT. Figure 7.11b shows some images that are not recognized by the oRGB-SIFT but are correctly recognized by the CSF. Figure 7.11c shows some images unrecognized by the CSF but are correctly

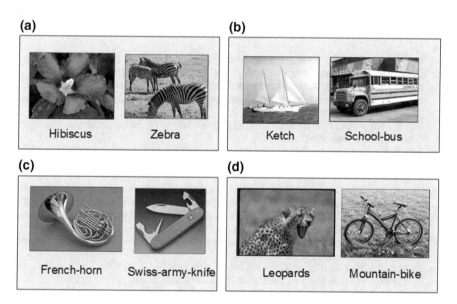

Fig. 7.11 Image recognition using the EFM-NN classifier on the Biometric 100 dataset: **a** example images unrecognized using the grayscale-SIFT descriptor but recognized using the oRGB-SIFT descriptor; **b** example images unrecognized using the oRGB-SIFT descriptor but recognized using the CSF descriptor; **c** images unrecognized using the CSF but recognized using the CGSF+PHOG. **d** Images unrecognized using the PCA but recognized using the EFM-NN on the CGSF+PHOG descriptor

recognized by the CGSF+PHOG descriptor. Lastly in Fig. 7.11d images unrecognized by the PCA but recognized by the EFM-NN classifier on the CGSF+PHOG descriptor.

7.9 Conclusion

In this chapter we presented the oRGB-SIFT feature descriptor and its integration with other color SIFT features to produce the Color SIFT Fusion (CSF), the Color Grayscale SIFT Fusion (CGSF), and the CGSF+PHOG descriptors. Experimental results using two large scale and challenging datasets show that our oRGB-SIFT descriptor improves the recognition performance upon other color SIFT descriptors, and the CSF, the CGSF, and the CGSF+PHOG descriptors perform better than the other color SIFT descriptors. The fusion of the Color SIFT descriptors (CSF) and the Color Grayscale SIFT descriptor (CGSF) show significant improvement in the classification performance, which indicates that the various color-SIFT descriptors and the grayscale-SIFT descriptor are not redundant for image classification.

References

1. Agarwal, S., Roth, D.: Learning a sparse representation for object detection. In: European Conference on Computer Vision, vol. 4, pp. 113–130. Copenhagen, Denmark (2002)
2. Banerji, S., A., V., Liu, C.: Novel color LBP descriptors for scene and image texture classification. In: 15th International Conference on Image Processing, Computer Vision, and Pattern Recognition, Las Vegas, Nevada (2011)
3. Bay, H., Tuytelaars, T., Van Gool, L.: SURF: Speeded up robust features. Comput. Vis. Image Underst. **110**(3), 346–359 (2008)
4. Bosch, A., Zisserman, A., Munoz, X.: Representing shape with a spatial pyramid kernel. In: International Conference on Image and Video Retrieval, pp. 401–408. Amsterdam, The Netherlands (2007)
5. Bosch, A., Zisserman, A., Munoz, X.: Scene classification using a hybrid generative/discriminative approach. IEEE Trans. Pattern Analy. Mach. Intell. **30**(4), 712–727 (2008)
6. Bratkova, M., Boulos, S., Shirley, P.: oRGB: a practical opponent color space for computer graphics. IEEE Comput. Graph. Appl. **29**(1), 42–55 (2009)
7. Burghouts, G., Geusebroek, J.M.: Performance evaluation of local color invariants. Comput. Vis. Image Underst. **113**, 48–62 (2009)
8. Csurka, G., Bray, C., Dance, C., Fan, L.: Visual categorization with bags of keypoints. In: Proceedings of Workshop Statistical Learning in Computer Vision, pp. 1–22 (2004)
9. Datta, R., Joshi, D., Li, J., Wang, J.: Image retrieval: ideas, influences, and trends of the new age. ACM Comput. Surv. **40**(2), 509–522 (2008)
10. Dobes, M., Martinek, J., Skoupil, D., Dobesova, Z., Pospisil, J.: Human eye localization using the modified hough transform. Optik **117**(10), 468–473 (2006)
11. Fergus, R., Perona, P., Zisserman, A.: Object class recognition by unsupervised scale-invariant learning. In: IEEE Conference on Computer Vision and Pattern Recognition, vol. 2, pp. 264–271. Madison, Wisconsin (2003)
12. Fukunaga, K.: Introduction to Statistical Pattern Recognition, 2nd edn. Academic Press, San Diego (1990)

13. Gevers, T., van de Weijer, J., Stokman, H.: Color feature detection: an overview. In: Lukac, R., Plataniotis, K. (eds.) Color Image Processing: Methods and Applications. CRC Press, University of Toronto, Ontario, Canada (2006)
14. Gonzalez, C., Woods, R.: Digital Image Processing. Prentice Hall, Upper Saddle River (2001)
15. Grauman, K., Darrell, T.: Pyramid match kernels: discriminative classification with sets of image features. In: International Conference on Computer Vision, vol. 2, pp. 1458–1465, Beijing (2005)
16. Griffin, G., Holub, A., Perona, P.: Caltech-256 object category dataset. Technical report, California Institute of Technology (2007)
17. Jurie, F., Triggs, B.: Creating efficient codebooks for visual recognition. In: International Conference on Computer Vision, pp. 604–610, Beijing (2005)
18. Ke, Y., Sukthankar, R.: PCA- SIFT: a more distinctive representation for local image descriptors. In: IEEE Conference on Computer Vision and Pattern Recognition, vol. 2, pp. 506–513, Washington, D.C. (2004)
19. Lazebnik, S., Schmid, C., Ponce, J.: A sparse texture representation using affine-invariant regions. In: IEEE Conference on Computer Vision and Pattern Recognition, vol. 2, pp. 319–324, Madison, Wisconsin (2003)
20. Lazebnik, S., Schmid, C., Ponce, J.: Semi-local affine parts for object recognition. In: British Machine Vision Conference, vol. 2, pp. 959–968, London (2004)
21. Leung, T., Malik, J.: Representing and recognizing the visual appearance of materials using three-dimensional textons. Int. J. Comput. Vis. **43**(1), 29–44 (2001)
22. Liu, C.: Capitalize on dimensionality increasing techniques for improving face recognition grand challenge performance. IEEE Trans. Pattern Anal. Mach. Intell. **28**(5), 725–737 (2006)
23. Liu, C.: Learning the uncorrelated, independent, and discriminating color spaces for face recognition. IEEE Trans. Inf. Forensics Secur. **3**(2), 213–222 (2008)
24. Liu, C., Wechsler, H.: Robust coding schemes for indexing and retrieval from large face databases. IEEE Trans. Image Process. **9**(1), 132–137 (2000)
25. Liu, C., Wechsler, H.: Gabor feature based classification using the enhanced Fisher linear discriminant model for face recognition. IEEE Trans. Image Process. **11**(4), 467–476 (2002)
26. Liu, C., Yang, J.: ICA color space for pattern recognition. IEEE Trans. Neural Netw. **2**(20), 248–257 (2009)
27. Lowe, D.: Distinctive image features from scale-invariant keypoints. Int. J. Comput. Vis. **60**(2), 91–110 (2004)
28. Mikolajczyk, K., Schmid, C.: A performance evaluation of local descriptors. IEEE Trans. Pattern Anal. Mach. Intell. **27**(10), 1615–1630 (2005)
29. Mikolajczyk, K., Tuytelaars, T., Schmid, C., Zisserman, A., Matas, J., Schaffalitzky, F., Kadir, T., Van Gool, L.: A comparison of affine region detectors. Int. J. Comput. Vis. **65**(1–2), 43–72 (2005)
30. Pontil, M., Verri, A.: Support vector machines for 3D object recognition. IEEE Trans. Pattern Anal. Mach. Intell. **20**(6), 637–646 (1998)
31. Schiele, B., Crowley, J.: Recognition without correspondence using multidimensional receptive field histograms. Int. J. Comput. Vis. **36**(1), 31–50 (2000)
32. Shih, P., Liu, C.: Comparative assessment of content-based face image retrieval in different color spaces. Int. J. Pattern Recog. Artif. Intell. **19**(7), 873–893 (2005)
33. Smith, A.: Color gamut transform pairs. Comput. Graph. **12**(3), 12–19 (1978)
34. Stokman, H., Gevers, T.: Selection and fusion of color models for image feature detection. IEEE Trans. Pattern Anal. Mach. Intell. **29**(3), 371–381 (2007)
35. Swain, M., Ballard, D.: Color indexing. Int. J. Comput. Vis. **7**(1), 11–32 (1991)
36. Verma, A., Liu, C.: Fusion of color SIFT features for image classification with applications to biometrics. In: 11th IAPR International Conference on Pattern Recognition and Information Processing, Minsk, Belarus (2011)
37. Verma, A., Liu, C.: Novel EFM- KNN classifier and a new color descriptor for image classification. In: 20th IEEE Wireless and Optical Communications Conference (Multimedia Services and Applications), Newark, New Jersey, USA (2011)

38. Verma, A., Liu, C., Jia, J.: New color SIFT descriptors for image classification with applications to biometrics. Int. J. Biometrics **1**(3), 56–75 (2011)
39. Verma, A., S., B., Liu, C.: A new color SIFT descriptor and methods for image category classification. In: International Congress on Computer Applications and Computational Science, pp. 819–822, Singapore (2010)
40. Weber, M., Welling, M., Perona, P.: Towards automatic discovery of object categories. In: IEEE Conference on Computer Vision and Pattern Recognition, vol. 2, pp. 2101–2109, Hilton Head, SC (2000)
41. Yang, J., Liu, C.: Color image discriminant models and algorithms for face recognition. IEEE Trans. Neural Netw. **19**(12), 2088–2098 (2008)
42. Zhang, J., Marszalek, M., Lazebnik, S., Schmid, C.: Local features and kernels for classification of texture and object categories: a comprehensive study. Int. J. Comput. Vis. **73**(2), 213–238 (2007)

Chapter 8
Clothing Analysis for Subject Identification and Retrieval

Emad Sami Jaha and Mark S. Nixon

Abstract Soft biometrics offer several advantages over traditional biometrics. With given poor quality data, as in surveillance footage, most traditional biometrics lose utility, whilst the majority of soft biometrics is still applicable. Amongst many of a person's descriptive features, clothing stands out as a predominant characteristic of their appearance. Clothing attributes can be effortlessly observable and described conventionally by accepted labels. Although there are many research studies on clothing attribute analysis, only few are concerned with analysing clothing attributes for biometric purposes. Hence, the use of clothing as a biometric for person identity deserves more research interest than it has yet received. This chapter provides extended analyses of soft clothing attributes and studies the clothing feature space via detailed analysis and empirical investigation of the capabilities of soft biometrics using clothing attributes in human identification and retrieval, leading to a perceptive guide for feature subset selection and enhanced performance. It also offers a methodology framework for soft clothing biometrics derivation and their performance evaluation.

8.1 Introduction

8.1.1 Motivations and Applications

Figure 8.1 shows a footage of rioters in London 2011 and it highlights a suspect with covered face and head. No soft traits are observable except clothing attributes. Also in the bottom right corner, a surveillance image released later within a list of most wanted suspects shows clearly what appears to be the same clothing, suggesting an

E.S. Jaha
King Abdulaziz University, Jeddah, Saudi Arabia
e-mail: ejaha@kau.edu.sa

M.S. Nixon (✉)
University of Southampton, Southampton, UK
e-mail: msn@ecs.soton.ac.uk

© Springer International Publishing AG 2017
C. Liu (ed.), *Recent Advances in Intelligent Image Search and Video Retrieval*,
Intelligent Systems Reference Library 121, DOI 10.1007/978-3-319-52081-0_8

Fig. 8.1 An image highlighting a suspect with covered face and distinct clothing, and *bottom right* a face image of a suspect appearing to wear the same clothes [37]

identification link to the rioter [37]. This image can provide a real example of how clothing attributes could be beneficial in identification and also demonstrates that, in some cases, clothing attributes can be the only observable soft traits to be exploited. In the worst case scenario, clothing can be used to narrow the search.

Clothing cannot only be considered to be individual, but can also reflect some cues regarding social status, lifestyle and cultural affiliation [9]. Clothing encodes more information about an individual, beyond just their visual appearance, which can reflect some cues regarding social status, lifestyle and cultural affiliation [81]. It is common for sections of societies to wear similar types of clothing, and some clothing styles appear to be correlated with age [31]. Youth cultures in particular are known to favour a particular sense of dress, such as punk, mod and goth, and there are many others. Clothing features may provide, in some sense, a number of indicators about potential behaviours or trends of a person. From the psychological perspective, someone's behaviour can be correlated with some clothing features and can be affected by certain clothing descriptions such as dark or black colour, as reported in some studies that wearing dark clothing could lead to more aggressiveness [81].

In daily life, people use clothing descriptions to identify or re-identify each other, especially from a distance or when the faces are not visible. Moreover, people can be unique by clothing and many people often wear similar clothing or a certain clothing style [37]. It has been shown that clothing attributes are naturally correlated and mutually dependent on each other [13]; this can be exploited when composing a biometric signature and even further to possibly infer some unknown attributes from the known ones.

The use of clothing is emphasized in the scope of person re-identification, which tends to be as one likely application can rely on soft clothing traits. It is rather that clothing is more suitable and effective for short term identification/re-identification, since people change clothes. Furthermore, in case of identification or re-identification at a distance, clothing attributes are most likely usable whilst the majority of traditional attributes and biometrics are disabled. Thus, biometric systems utilising soft clothing attributes can be widely deployed in forensic and security applications. Hence, the use of clothing for biometric purposes deserves more research interest than it has yet received.

8.1.2 Human Identification and Retrieval

Identifying people is an important task in everyday life. It is an urgent need and a routine task performed for different purposes mostly for security such as restricted access, border control and crime detection or prevention. The continuing technological advances continue to provide useful functional solutions for people identification. Many commonly-used solutions, for example username/password and smart cards, can suffer from violation and possible abuse [35].

Biometrics can solve these difficulties and provide effective automated solutions as they mostly use very discriminative traditional (hard) biometrics. Traditional biometric solutions exploit unique and inherent personal characteristics such as fingerprint, iris, and face (known as physical biometrics), or individual behaviours such as signature, voice, and gait (known as behavioural biometrics). Such classic physical and behavioural biometrics have been widely and effectively devoted to people identification and authentication.

In traditional biometric systems a person's traits need to be correctly enrolled into the database for successful use in biometric recognition or verification [61], which absolutely requires their cooperation. That is owing to the fact that such primary human traits can be influenced by many factors, such as distinctiveness, collectability, acceptability, universality, permanency and resistance to circumvention [39]. In addition, there are still several challenges and limitations to be confronted such as lower resolution and increased distance between the camera/sensor and the captured subject, where such hard traits lose utility [63]. Surveillance is a current example wherein the majority of hard biometrics, such as fingerprints, irises and maybe faces are impractical for identification.

Soft biometrics have recently emerged as a new attribute-based form of biometrics with a high level of usability and collectability offering many advantages over hard biometrics. Soft biometric traits are robust to viewpoint change and appearance variation [15]. In contrast with most hard traits, soft traits can be acquired using images/videos without a person's cooperation. Soft traits have been shown to address many problems and overcome limitations associated with hard traits.

Biometric based retrieval can be described as a task that aims to identify an unknown subject by comparing and matching their biometric signature with those signatures enrolled in a database [63]. The distinction of retrieval against identification that it concerns the ability to generalise to unseen data.

8.1.3 Soft Biometrics

Soft biometrics utilizes conventional human descriptions and translates these descriptions to the machine's biometric forms (in a way bridging the semantic gap) [68], thereby they can be used to characterize the person's face, body and accessories [20]. Soft biometric techniques mainly depend on defining a number of semantic attributes and assigning a set of descriptive labels (traits) for each attribute. Take for example, a semantic attribute "Height" which can be assigned a set of labels like ('Very short', 'Short', 'Average', 'Tall' and 'Very tall'). A semantic attribute can be any observable property that has a designated name or description by human like height, weight, gender, race, age, and eye or hair colour. Such attributes can be either binary attributes associated with categorical traits or relative attributes, associated with categorical or comparative traits. So far, the relative descriptions have been found to be more precise and informative compared with binary descriptions [59].

Despite of the lack of discriminatory capability associated with each single soft trait, its power can emerge when it is combined with other soft traits to be used as a biometric signature for identification [63], re-identification and database retrieval [55], or they can be used to augment other traditional physical and behavioural (hard) biometrics such as facial traits [56] or a gait signature [65, 68]. Soft traits could be automatically extracted and analysed then could be used to complement and assist some other primary traits in the recognition process [39]. Relative attributes are not only measurable representing the strength of attributes but also comparable allowing for more precise descriptions [59], which can be used to detect the small differences by describing the strength of attributes of a subject compared with others [66].

Soft body and face biometrics have been attracting increasing research interest and are often considered as major cues for identifying individuals [2–4, 53, 60, 63, 70, 78], especially in the absence of valid traditional hard biometrics. The basic approach uses human vision wherein labellers describe human body features using human understandable labels and measurements, which in turn allow for recognition and retrieval using only verbal descriptions as the sole query [67, 76]. The features also allow prediction of other measurements as they have been observed to be correlated [1]. Indeed, soft traits are not unique to an individual but a discriminative biometric signature can be designed from their aggregation. Since verbal identification can be used to retrieve subjects already enrolled in a database [70], it could be extended, in a more challenging application, for retrieval from video footage [63]. The capability of verbal retrieval from images and videos can pave the way for applications that can search surveillance data of a crime scene to match people to potential suspects described verbally by eyewitnesses. It is desirable for a multi-attribute retrieval approach to consider correlations and interdependencies among the query terms leading to refined search and enhanced retrieval results [74].

Soft biometric databases based on categorical labels can be incorporated with other biometrics to enhance recognition, such as integrating soft body traits with a gait signature [68] or soft face profile [90] with gait signature, and using soft facial traits along with other (hard) facial traits [56]. Nevertheless, soft comparative labels have been demonstrated to be more successful in representing the slight differences between people in body descriptions [53, 67] or can be also applicable on facial descriptions [2, 64].

In contrast to traditional face biometrics, facial soft biometrics are collectable in the absence of consent and cooperation of the surveillance subject, allowing for fast and enrolment-free biometric analysis [18]. Facial marks, for instance, can be automatically detected and described to be used as micro-soft traits to supplement primary facial features for improved face recognition and fast retrieval; besides, they may enable matching with low resolution or partial images [41, 60]. Measured facial information might be useful for gender prediction [22] and many system issues and challenges could arise when soft facial traits are used at a distance [3]. Face aging is a major challenge for any face recognition technique, and it becomes more challenging when added to unconstrained imaging (in the wild) even if it involves human intervention and decision [7] or further when addressing antispoofing in face recognition under disguise and makeup [83]. Therefore, soft facial attributes and features need to be carefully defined to be as much as possible describable and resistant against aging and other confounding factors, as such they can be more reliable and helpful for the purpose of subject identification in criminal investigations [42]. Complementing traditional face recognition systems by soft face biometrics is deemed to be a major research direction of several recent techniques [5] and a promising method significantly reducing recognition errors [86] and increasing the performance even under many complex conditions with a larger degree of variability in lighting, pose, expression, occlusion, face size and distance from the camera [54].

In hybrid biometric systems based on multiple primary traits such as faces and fingerprints, a number of soft traits can complement the identity information provided by the primary traits to improve the recognition performance [40]. For surveillance purposes, different forms of soft biometrics take place in a variety of applications and scenarios [15, 24–26, 34, 63, 78, 79, 84], where many of those soft traits could be easily distinguished even at a distance and then to be fused with vision features for target tracking [15] or to be fused with classic biometric traits, for the sake of improvement in overall recognition accuracy, especially with poor quality surveillance videos [78]. In surveillance images and videos, people's clothes are considered good discriminative features amongst the context-based information for distinguishing among people [88] and assisting identification [87]. A greater challenge for person re-identification exists across multi-camera surveillance, where typically assuming that individuals wear the same clothing between sightings, and likely some clothing attributes like colour and texture are observable with other soft body traits like height [8]. Note that age and sex are considered as important soft biometric traits for video surveillance [33]. Numerous recent approaches for people re-identification have been reviewed, discussed, and compared [80].

8.1.4   Clothing Information Analyses

8.1.4.1   Other Analyses of Clothing

Apart from biometric use, clothing details and features have been mostly used in automated search procedures, based on computer vision modelling or machine learning
techniques, through a variety of processes including detection, segmentation, feature extraction, categorization, clustering and retrieval [9, 11, 13, 16, 50, 51, 89].
With the continuing increase in research interest in clothing, a number of clothing
datasets, some of which are publically available, have been introduced and used
mostly for vision based clothing detection and recognition for purposes like classification in natural scenes [9] or search matching and retrieval [13, 50, 51]; other
approaches have been reviewed in [6]. Interestingly, a keen interest in semantic clothing description is reflected in the new and public Clothing Attribute Dataset [13].
APiS is a non-public database which has been designed to evaluate a classification
method for pedestrian multi-class attributes including few clothing descriptions such
as upper/lower clothing categories and colours [91]. Some other pedestrian datasets
like i LIDS and VIPeR have a variety of clothing representations that have been utilized to extract some clothing attributes for different analysis and purposes. PETA is
a recent pedestrian dataset designed with a larger scale and variability for pedestrian
attributes recognition at far distance [23].

 People choose and buy clothing via semantic attributes. Defining and utilizing
a list of clothing attributes for various purposes has been the concern of several
researchers [9, 13, 45, 50, 51, 85, 91]. Visual and semantic clothing attributes can
be used and applied in real-time frameworks with different aims including: semantic
clothing segmentation [16]; clothing fashion style recommendation via given user-
input occasion [50]; or by captured user-image used for matching and comparisons
[11, 89]; and online fashion search and retrieval [51]. Moreover, other than people
recognition or tag-and-track analysis, in surveillance videos, real-time visual clothing
segmentation and categorization can be performed, taking advantage of automatic
face detection to specify where clothing is likely to be located [85].

8.1.4.2   Clothing Analysis for Identity

There are few research studies associated with using clothing for biometric purposes [10, 14, 17, 21, 27, 47, 56, 75, 76, 79, 82, 84]. and just few of studies
associated with adopting and devoting a list of clothing attributes [45, 91], while
a number of most recent relevant and partly related approaches were reviewed in
[57] and [19] respectively. The majority of existing approaches employ computer
vision algorithms and machine learning techniques to extract and use visual clothing descriptions in applications including: online person recognition [56, 82]; soft
attributes for re-identification [45, 72] along with person detection [26, 34] and
tracking [14, 71] or attribute-based people search [26, 72]; detecting and analysing

semantic descriptions (labels) of clothing colours and types to supplement other body and facial soft attributes in automatic search and retrieval [9] or in automated person categorization [21]; utilising some clothing attributes like colour [75, 84] and style to improve the observation and retrieval at a distance in surveillance environments [76]; exploiting clothing colours and decorations to supplement other behavioural biometrics like human motion pattern, to form a biometric fingerprint that serves as a person's identifier [82] or other than to identity, to recognize gender via region-based clothing information in the case of insufficient face specification [10]; and appearance-based person re-identification in multi-camera surveillance systems [27]. There exists some work concerning semantic clothing attributes in fusion with vision features for surveillance purposes. Existing low-level features for person re-identification are complemented either by a full set of mid-level clothing attributes [47, 48] or by a selected and weighted subset of most-effective ones [45]. Whilst similar selection and weighting techniques enable optimised attributes [44] can be used in both automatic re-identification and identification [46].

8.1.4.3 Clothing as a Soft Biometric

It will be difficult to analyse clothing using computer vison in some surveillance images, given poor quality and low resolution, while human vision analysis offers supportive or alternative solutions [37]. In such surveillance images, whilst obtaining identifiable faces may be impossible, clothing appearance may become the main cue to identity [47]. Semantic clothing attributes can be naturally described by humans for operable and successful use in identification and re-identification [36–38]. Clothing is innately more efficient in short term id/re-id as people might change their clothes [38]. Even with images captured on different days, there remains sufficient information to compare and establish identity, since clothes are often re-worn or a particular individual may prefer a specific clothing style or colour [31]. Clothing descriptions such as indicative colour and decoration could be used to supplement other behavioural biometrics, like human motion patterns, hence they can form a biometric fingerprint that serves as a person's identifier [82] or other than identification, to recognize gender via region-based clothing info in the case of insufficient face specification [10].

People can be unique regarding their clothing and many people often wear similar clothing from day to day, or a certain clothing style [37]. Soft clothing traits are a new form of soft biometrics that can be associated with biometric signatures to achieve successful subject retrieval [36]. This motivates more interest in the latent ability of clothing information in subject identification and retrieval. Learning reliable biometric traits is required for realistic scenarios, such as after a change of viewpoint and partial occlusion [77]. In such scenarios, even some soft biometrics may likely be more vulnerable, especially to annotation subjectivity and missing information mostly caused by occlusion [36]. Viewpoint invariance is a challenging problem that has been considered in most biometric modalities [38]. For example a recognition algorithm suited for front-to-side re-identification, utilizing colour- and texture-based soft traits extracted from patches of hair, skin and clothes [17]. Subject retrieval is deemed as viewpoint invariant, if it remains invariant to any angle from which a subject is likely to be seen [32], such as front and side views.

8.1.5 Research Context

Although human clothes are a predominant visible characteristic of the person's appearance, they have yet to be adopted by the majority of research for representing soft biometric traits for an individual and have been considered unlikely as a cue to identity. Detecting the presence of some common clothing attributes, besides other soft biometrics, can supplement the low-level features used for person re-identification [45]. This allows use of more of the information available in surveillance video which is consistent with analysing data of such poor quality.

Other than biometrics, there has been previous work on recognizing clothing categories [51], semantically describing clothing pieces [13] and automatically detecting then classifying certain semantic clothing attributes in pedestrian data images [91], or alternatively classifying overall clothing styles in natural scenes based on a group of defined common categories and attributes [9]. Also automatic search and retrieval by clothing attributes for occasion-style recommendation [50] or a body attribute-based query integrated with clothing colours and types for a people search in surveillance data [79]. Comparative clothing attributes have been very rarely derived and used for biometric purposes or non-biometric like refining fashion search via comparative attributes adjusted by user feedback [43].

There is emergent work in using computer vision for automated clothing analysis of low- to medium-level features and very few high-level features (attributes). This is generally performed and could be practical on high resolution imagery with high contrast illumination and has yet to be efficiently and reliably applied to unconstrained surveillance imagery. Since this imagery is low resolution and poor quality, it will be difficult to analyse clothing information, especially if they are meant to be used as cues for identity. As such, attribute based approaches utilising human-vision, appear more suited to analysis of such surveillance imagery. In such imagery, whilst obtaining identifiable faces may be impossible, clothing appearance may become the main cue to identity [47]. Figure 8.2 demonstrates that this research context embraces clothing attributes, soft biometrics and human identification.

Even though clothing is innately more efficient in short-term identification/re-identification as people might change their clothes, the regions of clothing in images/videos can offer useful extra information about the identity of the individual. Moreover, for images captured on the same day, areas associated with clothing of these images may contain substantial information helping the discrimination between people, and even with images captured on different days, there remains sufficient information to compare and establish identity, since clothes are often re-worn or a particular individual may prefers a specific clothing style or colour [31].

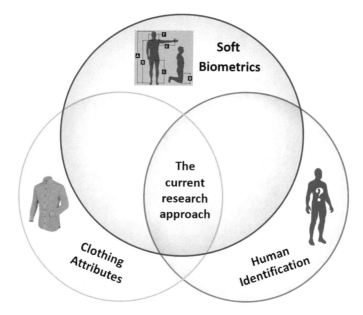

Fig. 8.2 The relevant areas of the current research approach

This chapter mainly provides an overview of clothing attributes for biometric purposes and an extended analysis with detailed representation of *soft clothing biometrics* and their capabilities in person identification and retrieval, which have been partly explained [37], analysed [36], and discussed for use in challenging scenarios [38] in our prior research introducing the first approach for identifying and retrieving people by their clothing as a major cue and a pure biometric. It also offers a methodology framework for soft clothing biometrics derivation and their performance evaluation. Furthermore, it studies the clothing feature space via performance analysis and empirical investigation of the capabilities of soft clothing biometrics in identification and retrieval, leading to an insightful guide for feature subset selection and enhanced performance.

8.2 Methodology

8.2.1 Experimental Framework of Soft Clothing Biometrics

We describe our exploratory experimental approach designed to obtain and analyse label and comparison data, which is used to build a soft biometric database of clothing descriptions and to investigate the significance and correlations of the proposed attributes, towards achieving and evaluating human identification and retrieval.

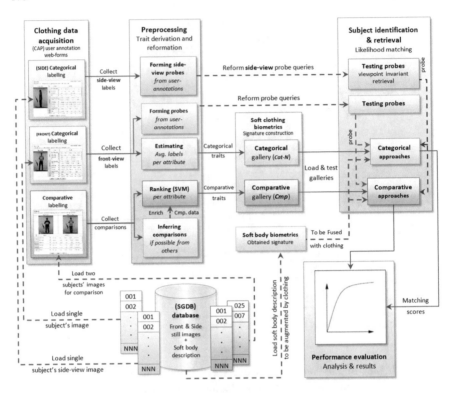

Fig. 8.3 Methodology framework of the current research approach

Figure 8.3 represents a high-level summary of our methodology used along this research, illustrated as an experimental framework.

This framework comprises four main stages. The first stage denotes clothing description data collection via Clothing Annotation Procedure website, where still subject images are loaded from Southampton gait dataset [73] and randomly displayed for human-based annotation. Two web-forms are associated with categorical labelling; one using a single front view image to be annotated for primary identification and the other using a single side view image to be annotated for more challenging retrieval tasks. A third web-form concerns pairwise comparative labelling for a subject against another.

The second stage shows pre-processing methods applied to collected labels and comparisons for trait derivation and feature vectors reformation for both gallery and probe sets. Thus a normalized average labels per attribute is estimated for each subject creating their biometric signature. Testing probes are randomly selected from available front and side categorical user annotations and reformed according to each examined gallery to be suited for matching and evaluation. For comparative clothing traits, attribute comparisons between subjects are firstly enriched by inferring all possible comparisons from available ones then the Ranking SVM method described in [37] is used to derive relative measurements to be used as comparative clothing traits. Consequently suitable testing probes are derived and mapped to corresponding relative measurements.

The third stage represents construction of all categorical and comparative galleries of soft biometric signatures used for identification and retrieval purposes. Finally the fourth stage outlines clothing-based subject identification and retrieval experiments using different proposed soft biometric approaches, leading to performance evaluation and analysis of soft clothing traits in supplementing soft body biometrics and further when used alone.

8.2.2 Soft Clothing Description Database

For clothing analysis and performance comparison we use our human-based soft clothing attributes for human identification [37] and subject retrieval [36, 38]. These soft biometrics were derived from clothing labels provided by a group of annotators for a proposed set of 21 soft clothing attributes, listed in Table 8.1 with their corresponding descriptive labels.

Seven of the 21 soft attributes (shown in bold) are derived using comparison such as (*Neckline size*: 'Much smaller', 'Smaller', 'Same', 'Larger', 'Much larger'). For non-relative soft attributes, each categorical label is assigned an integer value to represent the textual expression of the label. For better representation of relative soft attributes via categorical and comparative labels, we define bipolar scales in a way inspired from an early analysis that characterized human traits for whole-body descriptions [52]. Each of the seven relative soft attributes is formed as a categorical/comparative bipolar of a five-point scale.

8.2.3 Clothing Description Data Acquisition

Clothing annotations were obtained via a website and the annotation procedure comprises three tasks. The first task required a user to annotate front-view images of ten random subjects. Each subject was described by selecting 21 appropriate categorical labels. The second task required a user to compare one subject, selected randomly from the ten already annotated, with other ten new subjects. The third task is similar to the first task but it was designed to collect label data for side-view images instead of the front-view, as shown in Fig. 8.4, to be used only for retrieval performance analysis purposes.

All annotation web-forms were carefully designed to take into account a number of potential psychological factors, which were also considered earlier in collecting soft body labels [70], including:

Table 8.1 Attributes and their corresponding categorical and comparative labels

Body zone	Semantic attribute	Categorical labels	Comparative labels
Head	1. Head clothing category	[None, Hat, Scarf, Mask, Cap]	
	2. Head coverage	[None, Slight, Fair, Most, All]	[Much Less, Less, Same, More, Much more]
	3. Face covered	[Yes, No, Don't know]	[Much Less, Less, Same, More, Much more]
	4. Hat	[Yes, No, Don't know]	
Upper body	5. Upper body clothing category	[Jacket, Jumper, T-shirt, Shirt, Blouse, Sweater, Coat, Other]	
	6. Neckline shape	[Strapless, V-shape, Round, Shirt collar. Don't know]	
	7. Neckline size	[Very Small, Small, Medium, Large, Very Large]	[Much Smaller, Smaller, Same, Larger, Much Larger]
	8. Sleeve lengtli	[Very Short, Short, Medium, Long, Very Long]	[Much Shorter, Shorter, Same, Longer, Much Longer]
Lower body	9. Lower body clothing category	[Trouser, Skirt, Dress]	
	10. Shape (of lower clothing)	[Straight, Skinny, Wide, Tight, Loose]	
	11. Leg length (of lower clothing)	[Very Short, Short, Medium, Long, Very Long]	[Much Shorter, Shorter, Same, Longer, Much Longer]
	12. Belt presence	[Yes, No, Don't know]	
Foot	13. Shoes category	[Heels, Flip flops, Boot, Trainer, Shoe]	
	14. Heel level	[Flat/low, Medium, High, Very high]	[Much Lower, Lower, Same, Higher, Much higher]
Attached to body	15. Attached object category	[None, Bag, Gun, Object in hand, gloves]	
	16. Bag (size)	[None, Side-bag, Cross-bag, Handbag, Backpack, Satchel]	[Much Smaller, Smaller, Same, Larger, Much Larger]
	17. Gun	[Yes, No, Don't know]	
	18. Object in hand	[Yes, No, Don't know]	
	19. Gloves	[Yes, No, Don't know]	
General style	20. Style category	[Well-dressed, Business, Sporty, Fashionable, Casual, Nerd, Bibes, Hippy, Religious, Gangsta, Tramp, Other]	
Permanent	21.Tattoos	[Yes, No, Don't know]	

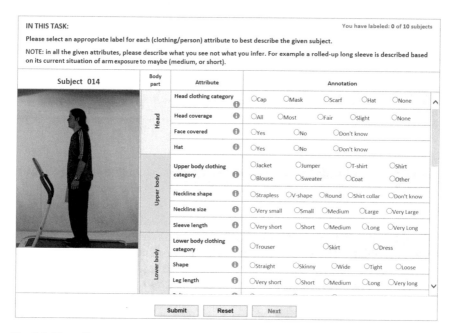

IN THIS TASK: You have labeled: 0 of 10 subjects

Please select an appropriate label for each (clothing/person) attribute to best describe the given subject.

NOTE: in all the given attributes, please describe what you see not what you infer. For example a rolled-up long sleeve is described based on its current situation of arm exposure to maybe (medium, or short).

Fig. 8.4 The online categorical annotation web-form of Task 3, describing side-viewpoint images

- *Anchoring*, which could occur when asking a question with some initial default answer a person's answer may often be anchored around those initial answers [12]. Therefore, to avoid this we left all labelling options (radio buttons) initially unchecked, forcing a pure response from the annotator.
- *Memory*, as the time passage could affect the recall of a subject's traits by a (witness) person [28]. Therefore, we allowed the annotator to see the described subject during the whole time of annotation to avoid the dependence on memory, assuring a confident description from the annotator to what is seen not what is remembered.
- *Observer variables*, which could affect a person's judgment of physical variables (measurements), since they often analyse other people in comparison with themselves, namely based on their perception to their own measurements [30, 58]. However this factor has less impact on clothing traits than body traits, since the annotator tends to focus more on clothes types and appearance. Therefore we emphasized repeatedly to the annotator that the objective is to describe a subject's clothing measurements relative to the subject's body (not the body itself), avoiding the annotator bias and confusion caused by the body measurements (e.g. confusing the trouser leg length with the subject leg length).
- *Defaulting*, which could occur when a person skips some descriptions since they assume them as known information by default [49]. Therefore, to avoid this problem we required responses to all questions as so the annotator could not leave out any description, forcing a response from the annotator for every single attribute

Table 8.2 Summary of obtained and inferred soft clothing data

Soft clothing data	Collected	Inferred	Total
Total user annotations	444	N/A	444
Total user comparisons	317	556	873
Total attribute annotations	9324	N/A	9324
Total attribute comparisons	2219	3892	6111

even if it is label could be known (by default) from those other attributes already given by the annotator. For example, if the upper clothing was labelled by the annotator as Jacket they may leave out the sleeve length, since it might be considered by their defaulting perception as long, while we still give some room of more discrimination (by specifying whether it is Long or Very long), and also allow for some possibility of exceptions (e.g. rolled up sleeve) or unusual fashions by giving the chance for the annotator to carefully choose the best descriptive label (which could be Short or Medium) based on what is exactly seen not what is inferred. Furthermore, we addressed some obvious potentials of annotators' defaulting by enforcing our controlled defaulting (i.e. applying auto label selection) in some definite cases when the selection of a particular label consequently and compulsory leads to another label (e.g. if head coverage labelled as None the hat presence definitely has to be No), which not only avoided the annotator's defaulting but also confliction between labels.

Thus, all front- and side-view samples of the 128 subjects in Southampton gait dataset were labelled by a group of 27 annotators and each subject was annotated and compared by multiple annotators, providing 21 categorical labels describing a single subject independently and seven comparative labels describing them with respect to another subject. From the available collected comparisons, we enrich comparative label data by inferring all possible comparisons, such that if subject i was compared with subjects j and k, we can infer the unavailable comparison between j and k. A summary of collected and inferred soft clothing data is shown in Table 8.2.

8.3 Analysing Soft Clothing Attributes

8.3.1 Attribute Significance and Correlation

The exploration of attributes' significance and correlations, is deemed to be an important analysis resulting in a better comprehension of which of attributes contribute most to identification and leading to wider potential predictability of other attributes [1]. The proposed clothing attributes were assessed to investigate correlation and effectiveness for subject description. Whilst different methods could be used to

investigate attribute significance, we used established statistical analysis for this
investigation. We used Pearson's r coefficient to compute the correlation and derive
the correlation matrix highlighting the significance of attributes and the mutual relations between attributes, which can be computed as:

$$r = \frac{\sigma_{XY}}{\sigma_X \sigma_Y} = \frac{\sum_{i=1}^{n}(X_i - \bar{X})(Y_i - \bar{Y})}{\sqrt{\sum_{i=1}^{n}(X_i - \bar{X})^2}\sqrt{\sum_{i=1}^{n}(Y_i - \bar{Y})^2}} \tag{8.1}$$

where X and Y are two variables representing the label values of two different
semantic traits used to describe an individual. Namely, X_i and Y_i are two labels
of ith annotation describing the same subject. So, r is calculated by dividing the
covariance of X and Y, denoted σ_{XY}, by the product of σ_X and σ_Y the standard
deviations of X and Y respectively. The resulting value of r ranges from -1 to 1
to represents the correlation coefficient between two labels. When the value of r is
1, it means that X and Y are positively correlated and their all data points lie on a
straight line. On the other hand, a value of $r = -1$, this implies that X and Y are
also correlated but negatively; $r = 0$ indicates that there is no correlation. Hence, the
closer r is to 1 reflects a higher positive relationship between labels. On the other
hand, the closer r is to -1 reflects a higher negative relationship between labels. Note
that, when one or both of correlated labels are binary or multi-class describing pure
(nominal) categorical attributes, this indicates that they are simultaneously present
in a single annotation. That is because the assigned numeric values of such labels do
not reflect ordering or measurements, unlike (ordinal) labels of relative attributes, so
they can be assigned values in any order or their values can be exchanged.

We generated several correlation matrices of all possible correlations for the available collected labels, comparisons and further derived traits. The calculated linear
correlation coefficient was considered as significant with respect to the resulting p-
value when $p \leq 0.05$. It is worth noting that the low correlation between two different
attributes does not suggest that there is no relationship between the two attributes,
but conveys a notion that this correlation is not prevalent within the clothing dataset
currently used such as the attributes 'Face covered' and 'Bag'.

8.3.1.1   Correlations Between Soft Clothing Labels

We study the proposed clothing attributes' relationships using the (Pearson's r) correlation matrix. We generated correlation matrices for the available collected labels,
comparisons and further derived traits. The calculated correlation coefficient was considered as significant when the resulting p-value when $p \leq 0.05$. Figure 8.5 demonstrates the correlation between all labels of 18 clothing attributes (see Table 8.1);

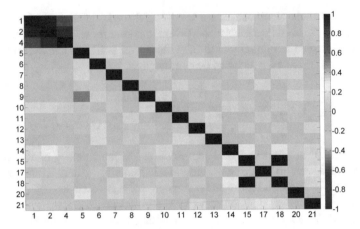

Fig. 8.5 Clothing label correlation matrix

traits without correlation are not shown. High correlation is symbolized by orange, and low by blue/green. As such, attributes relating to head coverage are highly correlated, as are the attributes (15) and (18) relating to the description of items attached to the body. Clothing is well correlated for upper (5) and lower (9) body. The matrix structure suggests that the desired uniqueness has been achieved.

8.3.1.2   Correlations Between Derived Soft Clothing Traits

The proposed categorical and comparative clothing traits were assessed to investigate their correlation and effectiveness for subject description. We generated correlation matrices for both categorical and comparative traits, which were derived from label data as normalized average labels and from comparison data as relative measurements, as described in Sect. 8.2.1. The calculated correlation coefficient was considered here also as significant when the resulting p-value when $p \leq 0.05$.

Figure 8.6 demonstrates the correlation between the most significant categorical and comparative traits (see Table 8.1); traits without correlation are not shown. High correlation is symbolized by orange, and low by blue/green. In the categorical matrix, traits relating to head coverage (2) and (4) are highly correlated, as are the traits (15) and (18) relating to the description of items attached to the body. Clothing categories are well correlated for upper (5) and lower (9) body, as expected. In the comparative matrix, neckline size (7) and heel level (14) are good correlated. Neckline size (7) is negatively correlated with leg length (11). The structures of both correlation matrices suggest that the desired uniqueness has indeed been achieved.

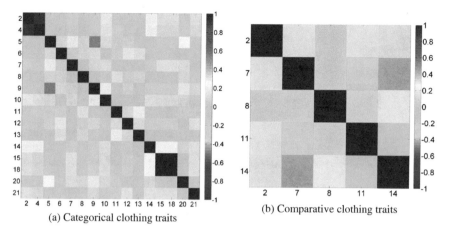

(a) Categorical clothing traits

(b) Comparative clothing traits

Fig. 8.6 Correlations between soft clothing traits

8.4   Analysing the Clothing Feature Space

8.4.1   Clothing Trait Distribution Statistics

Studying and analysing clothing trait distributions is useful not only to understand the nature of the population of these traits but also to enable comparison between clothing traits from a statistical point of view. Figures 8.7 and 8.8 demonstrate the population of each categorical and comparative soft clothing traits and clarify the differences in their data distributions. It can be perceived that some categorical traits such as '*Leg Length (of lower clothing)*' (11) and '*Neckline Size*' (7) appear to be more fitted to the normal distribution structure, whereas some others are less fitted such as '*Shoes category*' (13) and '*Shape (of lower clothing)*' (10).

   Thus, it is not necessarily for each clothing trait (in soft biometric signature) to be well fitted to normal distribution in order to be a valid soft biometric and some traits are expected to be more compliant to normal distribution than some others. This is in contrast to soft body traits, which are expected to offer more normal data distribution than clothing [69], since body attributes are natural and inherent, whereas clothing attributes are artificial and behavioural. This is due to the fact that populations of clothing attributes, (the presence/absence, and the prevalence/rarity) are more liable to many psychological, cultural, social, environmental, seasonal, and geographical factors, whilst this observation was quite similarly indicated in [10, 64]. For instance, a particular upper/lower clothing category can be predominant because of a certain seasonal factor like wearing coats in winter or shorts in summer. However, relative soft clothing traits appear to be less affected by such factors. For example, a relative attribute, describing sleeve/leg length will remain most likely operable, distinguishable, and comparable, even though for people wearing very

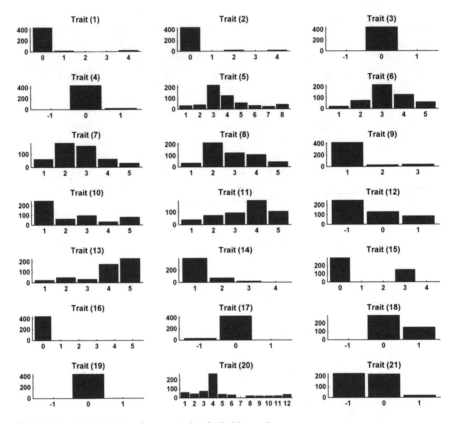

Fig. 8.7 Data distribution of categorical soft clothing traits

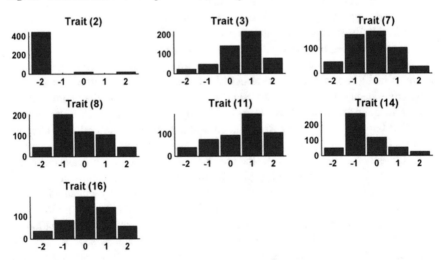

Fig. 8.8 Data distribution of comparative soft clothing traits

similar upper/lower clothes under the effect of any of those factors like people wearing suits in business or shirts in university.

Based on the obtained results of this analysis of soft clothing information, it is worth noting the following findings: regardless the type of soft clothing traits (categorical or comparative), the absence or rarity of a particular trait in a database will consequently affect the data distribution of that trait e.g. '*Head coverage*' (2) and '*Gloves*' (19); comparative traits often tend to display more normal distribution than their categorical counterparts, since they are represented by relative measurements derived using the comparative labels as explained previously in Sect. 8.2.2; and the categorical label data distributions of the binary/multi-class traits lean to be either unlikely or the least to satisfy the normal distribution assumption, as expected.

8.4.2 Analysis of Variance (ANOVA) of Clothing Traits

ANOVA is a widely used statistical test, and has been effectively used in several investigational studies. In our study, a one-way ANOVA is used to analyse the variance the variance for every single soft clothing trait, which is produced as a result of multiple annotation labels collected from multiple annotators describing the same subject.

The concept of this analysis process, known as the F-test, relies on analysing a single variable described by multiple observations, and these observations underlying two or more independent groups. This, in turn, allows comparison between intra-variance *within* each of all groups with the mean inter-variance *between* these groups, resulting in a value defined as F-ratio. A further value can be deduced form F-ratio and the F-distribution known as p-value, which is supposed to be less than or equal to a certain small value (usually 0.05 or 0.01), to consider F-value as significant and to reject the null-hypothesis [29].

The computed F-ratio and its corresponding p-value are used to order the categorical and comparative soft clothing biometrics by their estimated capability of distinguishing between subjects. A larger F-ratio with a smaller p-value is better, where the total degree of freedom $(df) = 443$ for F-ratio, and $p \leq 0.05$ for significance level. F-ratio is calculated as follows:

$$F\text{-}ratio = \frac{\sum_i (\bar{X}_i - \bar{X})^2/(K-1)}{\sum_{ij}(X_{ij} - \bar{X}_i)^2/(N-K)} \tag{8.2}$$

The upper bound in this Eq. (8.2) yields the (total between-groups variance), whereas the lower bound yields the (total within-group variance). In our biometric context, each group represents a subject, where K is the number of subjects and N the total number of samples of all subjects, while n_i represents the number of sample of ith subject. X_{ij} denotes the value of jth sample of the ith subject, $\bar{X}_i$ denotes the mean across all samples of the ith subject, and $\bar{X}$ is the mean across all N samples.

Table 8.3 Ordered list of clothing traits by their F-ratios

Clothing attribute		F-ratio ($df = 448$)	p-value (p $\leq$ 0.05)
(a) Categorical traits			
9.	Lower body clothing category	19.747	2.02E-99
2.	Head coverage	19.651	3.90E-99
8.	Sleeve length	12.590	3.60E-74
1.	Head clothing category	11.054	2.46E-67
11.	Leg length	5.040	5.48E-32
4.	Hat	4.913	4.87E-31
5.	Upper body clothing category	4.738	1.03E-29
12.	Belt presence	4.457	1.52E-27
13.	Shoes category	4.282	3.64E-26
6.	Neckline shape	2.413	1.68E-10
20.	Style category	2.349	5.83E-10
14.	Heel level	1.863	5.58E-06
7.	Neckline size	1.688	1.15E-04
10.	Shape	1.626	3.20E-04
17.	Gun	0.862	8.34E-01
21.	Tattoos	0.821	9.02E-01
15.	Attached object category	0.721	9.84E-01
18.	Object in hand	0.721	9.84E-01
3.	Face covered	n/a	n/a
16.	Bag	n/a	n/a
19.	Gloves	n/a	n/a
(b) Comparative traits			
2.	Head coverage	15.670	3.68E-86
8.	Sleeve length	12.021	1.02E-71
11.	Leg length	4.819	2.49E-30
16.	Bag	4.689	2.42E-29
3.	Face covered	2.559	9.38E-12
7.	Neckline size	2.515	2.26E-11
14.	Heel level	1.636	2.73E-04

The degree of freedom is enforced by $K-1$ on subject level and $N-K$ on sample level.

The reported results show the most significant soft clothing traits, and likely suggest which traits are more successful than others. Table 8.3 provides the ordered lists of resulting ANOVA test values for categorical and comparative clothing traits. Lower body clothing category (9) is observed as the most discriminative trait. Head coverage (2) is highly discriminative since few subjects had covered heads. It is perhaps surprising that sleeve length (8) is so discriminative, especially compared

Table 8.4 MANOVA of soft clothing traits showing significance by different standard statistics

MANOVA statistic	(a) Categorical traits		(b) Comparative traits	
	F-ratio (df = 443)	p-value (p ≤ 0.05)	F-ratio (df = 443)	p-value (p ≤ 0.05)
Pillai's Trace	2.314	<0.0001	3.343	<0.0001
Wilks' Lambda	3.314	<0.0001	4.729	<0.0001
Hotelling's Trace	5.186	<0.0001	6.853	<0.0001
Roy's Largest Root	23.519	<0.0001	20.045	<0.0001

with the length of the trousers (11), but that is what this analysis reveals, and no summary analysis is possible by human vision. It is worth applying further multivariate analysis of variance (MANOVA) on the whole set of clothing attributes, which explores interactions between multiple variables (attributes) and measures significance of multivariate discrimination capability via different standard statistics [29], as shown in Table 8.4, where all measures provide significant p-value <0.0001.

8.4.3 Mutual Dependence Analysis of Clothing Traits

8.4.3.1 Statistical Dependency

Statistical Dependency (SD) is a method used to examine whether the values of a feature are dependent on the associated class labels and whether those feature values and class labels are unlikely to co-occur by chance [62]. Here in our biometric context, the feature is a soft clothing trait and the class labels representing subject IDs. The larger the SD value, the higher is the dependency between trait values and subject IDs. While when a trait is fully independent of the subject IDs, it receives a minimal value of SD ≤ 1 and reflects insignificant quantity of dependence. The SD measure is calculated for each soft clothing trait and can be defined as:

$$SD = \sum_{x \in X} \sum_{y \in Y} p(x, y) \frac{p(x, y)}{p(x)p(y)} \tag{8.3}$$

where X a variable denoting a trait value, and Y a variable representing the class label (i.e. the subject's identity). $p(x, y)$ is the joint probability density function of the random variables X and Y, where $p(x)$ and $p(y)$ are the marginal probability density functions respectively. Table 8.5 shows categorical and comparative soft clothing traits ordered by the SD measure. It appears natural that the comparative traits are more statistically dependent by receiving higher SD values than the categorical traits. Upper body clothing category (5) appears as the best categorical trait in terms of SD with a far high score, followed next by neckline shape (6) and style category (20) both

Table 8.5 Statistical dependency (SD) analysis of soft clothing traits

Clothing attribute		SD
(a) Categorical traits		
5.	Upper body clothing category	2.483
6.	Neckline shape	1.949
20.	Style category	1.946
8.	Sleeve length	1.854
10.	Shape (of lower clothing)	1.765
7.	Neckline size	1.732
11.	Leg length (of lower clothing)	1.504
13.	Shoes category	1.409
12.	Belt presence	1.323
1.	Head clothing category	1.195
2.	Head coverage	1.195
9.	Lower body clothing category	1.094
14.	Heel level	0.926
4.	Hat	0.861
21.	Tattoos	0.757
15.	Attached object category	0.350
18.	Object in hand	0.350
17.	Gun	0.018
3.	Face covered	1.29E-04
16.	Bag	1.29E-04
19.	Gloves	1.29E-04
(b) Comparative traits		
8.	Sleeve length	4.524
2.	Head coverage	4.310
16.	Bag (size)	4.125
3.	Face covered	3.927
7.	Neckline size	3.813
14.	Heel level	3.765
11.	Leg length	3.463

with almost the same score. Sleeve length (8) is the highest dependent comparative trait and among the highest categorical traits, which may emphasize the usefulness of this trait.

8.4.3.2 Mutual Information

Mutual information (MI) is a method used to measure the mutual dependence of two variables. Note that the concept and the way of calculation of MI is very similar to the statistical dependency (SD) explained in the previous section. It is worth noting that analysing mutual dependence among soft clothing attributes via SD or MI has not been applied prior to this research. SD may be deemed more sensitive and preferable in some high-dimensional classification problem like voice analysis [62]. Nevertheless, MI is more common in use and meant to provide more accurate translation of mutual dependence between variables by enforcing the logarithmic compression shown in Eq. (8.4).

$$MI = I(X, Y) = \sum_{x \in X} \sum_{y \in Y} p(x, y) \log \left(\frac{p(x, y)}{p(x)p(y)} \right) \tag{8.4}$$

The MI quantity (usually denoted $I(X, Y)$) is computed here in two ways: (i) similar to the way used to compute SD (Sect. 8.4.3.1), considering X as a variable denoting a trait value, and Y a variable representing the subject's identity; and (ii) considering both X and Y as trait values of two different soft clothing traits, resulting in measuring MI of these two traits. $p(x, y)$ is the joint probability density function of the random variables X and Y, where $p(x)$ and $p(y)$ are the marginal probability respectively. Table 8.6 shows categorical and comparative soft clothing traits ordered by MI deduced (based on aspect (i)) for each trait against the subject IDs. While Fig. 8.9 shows a symmetric color-coded matrix of mutual information between each two soft clothing traits (based on aspect (ii)).

It can be observed that analysis of MI in Table 8.6 presents different ranking of clothing traits compared with the trait ranking presented by SD in Table 8.5, suggesting that MI and SD differently measure the mutual dependence, even though they are based on the same concept. Upper body clothing category (5), neckline shape (6), and sleeve length (8) remain on top of the rank unlike style category (20) which fall back in rank. Categorical leg length (11) interestingly becomes the next high mutually informative trait after sleeve length (8), whereas it is the lowest among the comparative traits.

In Fig. 8.9, normalized MI measurements (ranging from 0 to 1) of all possible pairs of soft clothing traits are visually illustrated, where bright colours (white, yellow) reflect high MI values, vivid colours (orange, red) display medium values, and dark colours (brown, black) represent low values. The diagonal quantities are all set to zero (black), enforcing the notion that a variable cannot reduce the uncertainty of itself. As such, in the categorical matrix, it can be perceived that the traits (15) and (18) relating to the description of objects attached to the body are the highest mutually informative traits. Most observable MI values are located between nine traits (5–13), where each provides some mutual information with the other eight, ranging from low to medium. Upper clothing category (5) shows well mutual information with neckline shape (6) and sleeve length (8), signifying the compatibility between a certain upper clothing type and the specification of its parts (neckline, sleeve etc.). Clothing style category (20) as a general description offers some notable mutual information with upper and lower body clothing traits (from 5 to 13). This intuitively reflects the reality that detailed characteristics of upper and lower clothing explicitly contribute to inform

Table 8.6 Mutual information (MI) analysis of soft clothing traits

Clothing attribute		MI $I(X_i, Y)$
(a) Categorical traits		
5.	Upper body clothing category	1.387
6.	Neckline shape	1.289
8.	Sleeve length	1.130
11.	Leg length (of lower clothing)	0.993
13.	Shoes category	0.976
20.	Style category	0.972
10.	Shape (of lower clothing)	0.946
7.	Neckline size	0.843
12.	Belt presence	0.815
9.	Lower body clothing category	0.329
14.	Heel level	0.293
21.	Tattoos	0.246
15.	Attached object category	0.202
18.	Object in hand	0.202
1.	Head clothing category	0.067
2.	Head coverage	0.067
4.	Hat	0.045
17.	Gun	0.001
3.	Face covered	1.44E-06
16.	Bag	1.44E-06
19.	Gloves	1.44E-06
(b) Comparative traits		
8.	Sleeve length	2.288
2.	Head coverage	2.242
16.	Bag (size)	2.184
3.	Face covered	2.116
14.	Heel level	2.076
7.	Neckline size	2.042
11.	Leg length	1.836

what clothing style is worn by a person, and the way around, a particular clothing style can give an idea about likely upper and lower clothing descriptions. On the other hand, in the comparative matrix, the highest MI quantity is between neckline size (7) and sleeve length (8). The next significant mutual information exists between neckline (7) and bag (16) sizes, and sleeve (8) and leg (11) lengths.

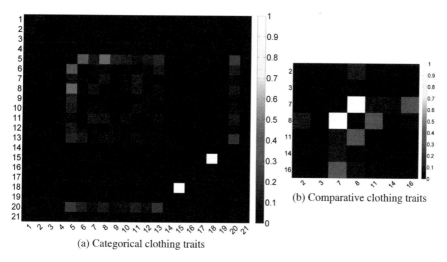

(a) Categorical clothing traits

(b) Comparative clothing traits

Fig. 8.9 Mutual information matrix of soft clothing traits

8.4.4 Evaluating Clothing Features by Performance

In this analysis, we utilize five different methods to evaluate then rank or list in order the manual/automatic soft clothing features by either their ranking scores or suggested feature selection ordering. This analysis leads to better empirical understanding of how significant and relevant are those clothing features to the human identification problem. Moreover, the investigation of feature significance and relevance can be used further to guide the feature subset selection, achieving the highest possible performance with a minimum number of features. This analysis is applied to the main two soft clothing forms as follows:

- Categorical soft clothing traits referred to as (*Cat-21*).
- Comparative soft clothing traits referred to as (*Cmp-7*).

For this analysis, the adopted scoring methods comprises: ANOVA, SD, and MI which are explained in Sects. 8.4.2 and 8.4.3, where the resulting scores are used to rank clothing traits per each method. In addition, two more feature subset selection methods are used to order features by their cooperative usefulness in recognition based on the way of ordering (or prioritising in selection) enforced by inclusion/exclusion steps per each method. These two methods are Sequential Forward Selection (SFS) and Sequential Floating Forward Selection (SFFS).

Throughout this chapter, 'T' denotes the *trait* of clothing followed by its number (shown in Table 8.1) to make them clearer across different orderings. Besides, all experiments for clothing trait performance evaluation are achieved using the challenging methodology of viewpoint invariant retrieval described in [38].

Recognition rates over the feature subset size increase are depicted for all approaches at three ranks: rank 1 (the best match at the top rank); rank 10 (the capability of correct retrieval within a list of top ten matches); and rank 128 (the average accuracy over all matches at the full-rank). Standard Error of the Mean (SEM) associated with estimating an average-score are shown for rank 10 and rank 128 and computed as $\pm\text{SEM} = \sigma/\sqrt{k}$ where σ is the standard deviation of all scores up to rank k.

In the Figs. 8.10 and 8.14, which show and compare the ranking scores of all traits ordered by the average of normalized scores for *Cat-21*, and *Cmp-7* respectively. The corresponding scores of all traits are normalized per each scoring method (i.e. ANOVA, SD, and MI) using min-max normalization to rescale all values into the range [0, 1]. As such, given S as a set of all scores inferred by a single method, for each score $s \in S$, a new normalized score is deduced as $s' = (s - min_S)/(max_S - min_S)$, where min_S is the minimum score in S and max_S is the maximum score in S.

SFS and SFFS are well known and widely used feature subsect selection methods in practice [62]. They both work in similar way by starting with an empty set then iteratively include the next best trait to the subset of selection until stopping when the highest possible performance is reached. The main difference is that SFFS is a (wrapper) method that performs some possible backward exclusion steps (of the worst traits) after each forward inclusion step (of the best trait), as long as the new subset (after the exclusion) increases the previous performance of a subset with the same size. This is what is known as floating exclusion step, which is applied by the Sequential Backward Selection (SBS) algorithm.

Unlike ANOVA, SD, and IM scoring methods, SFS and SFFS do not enforce ranking or scoring for each trait independently. It is rather that they enforce prioritising (in incrementally selecting) each trait in relation to the other traits in the selected subset. We utilize SFS and SFFS algorithms further for sorting traits based on a sequence suggested by either algorithm, in which a trait is the best as a next selection to be added to a (cumulative) subset achieving the highest (incremental) recognition score. As such, with respect to the ordering suggested by SFFS algorithm (see Table 8.7), a sequence of traits comprising M20, M11, M8 does not necessarily mean that M20 is the best among these three traits when used alone, nor that M11 is worse than M20 or better than M8 in performance. It can rather mean that a combination of M20, M11 achieves better performance than a combination of M11, M8 and a further combination of all M20, M11, M8 is better than a combination of any two of them. This example is applicable on any three sequential traits suggested by SFS such as M5, M20, M11, in Table 8.7, except that only M5 is indeed the best trait in performance among all traits when used alone (i.e. feature subset size = 1), since the first step of SFS algorithm tests each trait individually to find out which feature achieves the highest performance to be included in the feature subset and it is impossible to be excluded as in SFFS algorithm, where all selections even the first may be excluded from the feature subset for one or more times.

Note that, to obtain the SFS and SFFS ordering lists shown in the Tables 8.7 and 8.8, we proceed to include all traits rather than terminating when only the best performance is achieved with reduced dimensionality. This is because we aim to prioritise all traits not just a subset of the best trait as in the standard use but even the ones may do not change or lower the performance, and to suggest ordering for the incremental feature subset selection starting from one trait and ending with the complete set of traits.

8.4.4.1 Evaluating Categorical Clothing Features

Table 8.7 shows the ordering lists of categorical soft clothing traits (*Cat-21*), which are deduced using SFS and SFFS methods. It can be observed that each method suggests a different ordering for the examined traits. Upper clothing category (T5) is reported earlier (in Sects. 8.4.2 and 8.4.3) to be the best trait by SD and MI, while it is not found as the best discriminative trait by ANOVA. T5 is also listed on top of SFS ordering, whereas it is dropped down the ordering list suggested by SFFS in Table 8.7. On the other hand, Lower body category (T9) is granted rank 1 only once by ANOVA and Style category (T20) is suggested as the first trait in SFFS ordering. In overview, Sleeve length (T8) appears to be the most useful trait, since it continuously remains at either rank 3 or 4 with respect to all rankings, also it is the

Table 8.7 Ordering categorical clothing traits *Cat-21* using SFS and SFFS

Cat-21 Soft clothing traits		
Ordering	SFS	SFFS
1	T5	T20
2	T20	T11
3	T11	T8
4	T8	T10
5	T10	T13
6	T13	T6
7	T6	T12
8	T7	T14
9	T1	T9
10	T12	T2
11	T15	T3
12	T14	T4
13	T9	T16
14	T2	T19
15	T3	T17
16	T4	T15
17	T16	T18
18	T18	T21
19	T19	T5
20	T21	T1
21	T17	T7

Table 8.8 Ordering comparative clothing traits *Cmp-7* using SFS and SFFS

Cmp-7 Soft clothing traits		
Ordering	SFS	SFFS
1	T11	T11
2	T2	T8
3	T8	T14
4	T3	T7
5	T16	T2
6	T14	T3
7	T7	T16

Fig. 8.10 *Cat-21* traits ordered by average min-max normalized ranking scores to the range [0, 1]

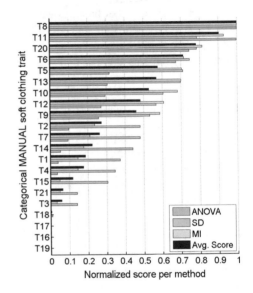

third in SFFS ordering and the fourth in SFS ordering. Leg length (T11) is generally the next in usefulness, which appears within the range from 2 to 7 throughout all rankings and ordering lists. Figure 8.10 shows and compares the ranking scores of *Cat-21* traits ordered by the average of min-max normalised scores per method to the range [0, 1]. It can be seen that the four top traits (T8, T11, T20 and T6) indeed receive far higher scores by all ranking methods.

Figures 8.11, 8.12, and 8.13 show recognition rates at rank 1, 10, and 128 respectively using *Cat-21* clothing traits, which enable performance comparison between the five rankings or ordering along all possible sizes of the feature subset (i.e. from 1 to 21). SFS attains the best performance in average and ANOVA reaches the highest possible performance using only 11 traits, as shown in Fig. 8.11. MI achieves the highest average performance overall subset sizes from one to 21 as in as in Figs. 8.12

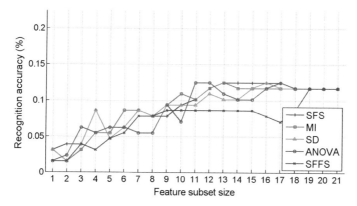

Fig. 8.11   Recognition rates at **rank 1** of *Cat-21* per method along the feature subset size increase

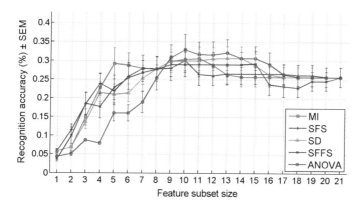

Fig. 8.12   Recognition rates at **rank 10** of *Cat-21* per method along the feature subset size increase

and 8.13 respectively, while here again ANOVA obtains the best possible recognition rate at both 10 and 128 ranks using only ten traits.

8.4.4.2   Evaluating Comparative Clothing Features

Table 8.8 shows the ordering lists of comparative clothing traits (*Cmp-7*) deduced using SFS and SFFS methods. In overview, Sleeve length (T8) remains here as the most powerful trait and occupies either rank 1 or 2 in the three rankings (see Sects. 8.4.2 and 8.4.3). T8 is also suggested as either the second or third trait in SFFS and SFS orderings respectively as in Table 8.8. The three traits T8, Head coverage (T2), and Leg length (T11) are shown as the most useful features in overall rankings and orderings. T2 is the top trait by ANOVA, T8 is the supreme by SD and MI. T11 is in the top of SFS and SFFS ordering lists. Figure 8.14 shows min-max normalised ranking scores per method for *Cmp-7* traits sorted by average score. It shows T8 as

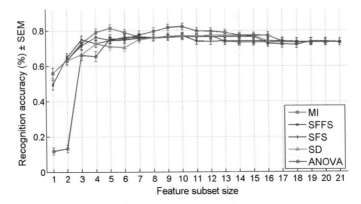

Fig. 8.13 Recognition rates at **rank 128** of *Cat-21* per method along the feature subset size increase

Fig. 8.14 *Cmp-7* traits
ordered by average min-max
normalized ranking scores to
the range [0, 1]

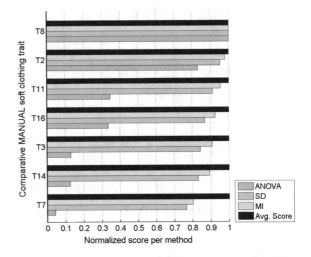

the most powerful trait followed by T2 as they receive the highest ranking scores by
all means.

Recognition rates using *Cmp-7* traits are deduced at rank 1, 10, and 128 and presented in Figs. 8.15, 8.16 and 8.17 respectively, comparing the performance between
the five ranking or ordering methods along all possible sizes of a feature subset (i.e.
from 1 to 7). In Fig. 8.15 it is clear that all methods achieve very slight improvement
in the recognition rate at rank 1 along the subset size increase from one to seven,
as expected, where the best possible performance is obtained by both SD and MI
using a subset consisting of only six traits and they both in addition to SFFS receive
the highest average recognition overall subset sizes. In Fig. 8.16 the highest average
performance and the top recognition rate using only five traits are scored by SFFS
followed by ANOVA. At rank 128, as shown in Fig. 8.17, SFFS and ANOVA equally
receive the highest possible performance with only six traits but ANOVA gains the

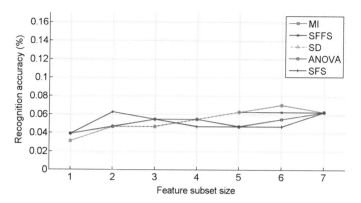

Fig. 8.15 Recognition rates at **rank 1** of *Cmp-7* per method along the feature subset size increase

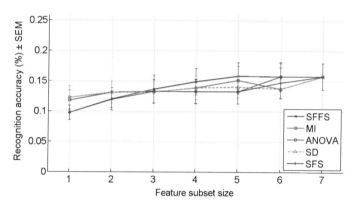

Fig. 8.16 Recognition rates at **rank 10** of *Cmp-7* per method along the feature subset size increase

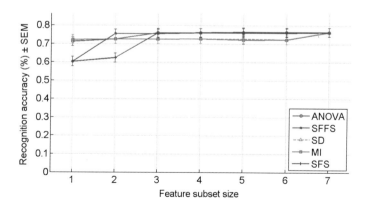

Fig. 8.17 Recognition rates at **rank 128** of *Cmp-7* per method along the feature subset size increase

best average performance overall subset sizes. In general, it can be perceived that these comparative traits contribute somewhat similarly in the recognition. Therefore, having them in different rankings or orderings and using different subsets with the same size is not expected to considerably enhance the performance.

8.5 Experiments

Here a number of experimental examples are presented, showing how soft biometric identification and retrieval can be achieved in different aspects when clothing attributes are used alone or in fusion with other traditional soft biometrics like body attributes.

8.5.1 Identification Using Soft Clothing Biometrics

Each of the proposed approaches (*Cat-21*, *Cat-7* and *Cmp*) is used in isolation for human identification, when soft clothing traits are the only used biometric. This can allow investigation of the pure discrimination power of soft clothing traits. Table 8.9 lists and describes the soft biometric approaches (and produced galleries) used in this experiment.

Here the leave-one-out method is also applied to probe the three soft clothing galleries using the appropriate probe set for each. The CMC scores and the ROC analysis of the proposed approaches are given in Table 8.10. Although all the approaches tend to improve the identification score sharply throughout the rank increase from 1 to 10, the categorical labels *Cat-21* start from a much better score and gain a much higher average score than *Cat-7* and *Cmp* as shown in Fig. 8.19 comparing their CMC performance. The *Cat-21* approach outperforms the other approaches in all terms, but the decidability d' is the largest and best in *Cmp*. It is noteworthy that unlike in augmenting soft biometrics, the comparative traits *Cmp* achieve a better performance than their categorical counterparts underlying *Cat-7*. Also in terms of

Table 8.9 Soft clothing approaches used for identification performance evaluation

Approach	Description
Cat-21	21 categorical normalized average-labels, describing 14 binary/multiclass and 7 relative attributes
Cat-7	categorical normalized average-labels, a subset of Cat-21 describing only the 7 relative attributes
Cmp	7 comparative traits, mapping Cat-7 to corresponding relative measurements based on comparisons per each of the 7 relative attributes

Table 8.10 CMC and ROC scores of soft clothing biometrics when used alone for identification

Approach	Top rank	AVG sum match scores up to rank		100% accuracy achieved at rank	EER	AUC	d'	Overall rank
	= 1	= 10	= 128					
Cat-21	**0.63**	**0.843**	**0.984**	**41**	**0.137**	**0.059**	1.442	**1**
Cat-7	0.27	0.507	0.923	92	0.192	0.108	1.303	3
Cmp	0.28	0.510	0.929	96	0.174	0.088	**1.824**	2

Table 8.11 Soft body and clothing approaches (galleries) for retrieval experiments

Body-based soft biometrics	
tradSoft	4 categorical soft body biometrics (Age, Ethnicity, Sex, and Skin Colour)
softBody	17 categorical soft body biometrics including tradSoft (observable from front/side views)
Combined soft clothing & body biometrics	
tradCat-21	21 categorical clothing traits combined with *tradSoft*
softCat-21	21 categorical clothing traits combined with *softBody*
tradCat-6	The best 6 categorical clothing traits with *tradSoft*
softCat-6	The best 6 categorical clothing traits with *softBody*
tradCmp	7 comparative clothing traits combined with *tradSoft*
softCmp	7 comparative clothing traits combined with *softBody*

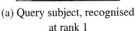

(a) Query subject, recognised   (b) Left: Query subject, recognized at rank 65.
at rank 1   Right: confused closest match at rank 1

Fig. 8.18 Matching performance by clothing alone: **a** good match, and **b** poor match

ROC performance, Fig. 8.20 illustrates that *Cmp* accordingly received less error than *Cat-7*.

Figure 8.18 shows a good and poor samples of identification by clothing, where (a) a subject successfully recognised by given query annotations at rank 1, whereas (b) by given query annotations describing the left subject, he is poorly recognized at rank 65, while the right subject was incorrectly the closest match. *Cat-21* provides the smallest EER in identification. Here, *Cmp* compared with *Cat-7* achieves slightly better identification and receives relatively larger EER. That is obviously consistent with numerical scores in Table 8.12.

8.5.2 Retrieval by Soft Clothing Biometrics

In this experiment we use the proposed approaches in challenging biometric retrieval to investigate the capability of soft clothing traits when they are generalised to unseen

Fig. 8.19 CMC performance (up to rank 100) of soft clothing biometrics when used alone

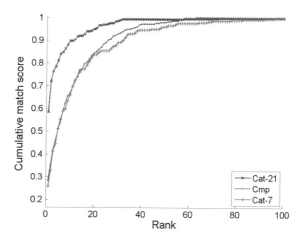

Fig. 8.20 ROC performance of soft clothing biometrics when used alone

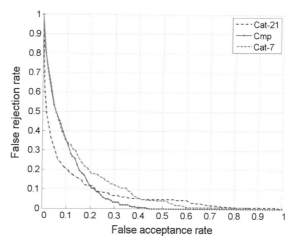

samples and unknown description data. We also aim to explore their performance in fusion and their capabilities to supplement other soft biometrics. Table 8.11 describes the produced soft biometric approaches (and produced galleries) of body traits and when combined with soft clothing traits.

8.5.2.1 Retrieval by Soft Clothing and Traditional Soft Biometrics

Here, clothing descriptions are used to supplement the traditional soft traits (Age, Ethnicity, Sex, and Skin Colour) in subject retrieval (see Table 8.11), the experimental results detailed in Table 8.12 and the CMC curves shown in Fig. 8.21 suggest that the retrieval performance is considerably and consistently enhanced by clothing traits in *tradCmp* and *tradCat-6* of clothing approaches. The retrieval accuracy of *tradSoft*

Table 8.12 Performance metrics of traditional soft biometrics and when add soft clothing traits

Approach	Top rank = 1	AVG sum match scores up to rank		100% accuracy achieved at rank	EER	AUC	d'	Overall rank
		= 10	= 128					
tradSoft	0.176	0.347	0.872	73	0.183	0.127	1.882	3
tradCat-21	0.234	0.443	0.865	104	0.302	0.198	0.882	4
tradCat-6	**0.346**	**0.617**	0.944	**58**	0.144	0.090	1.223	2
tradCmp	0.308	0.607	**0.946**	69	**0.113**	**0.077**	**2.006**	**1**

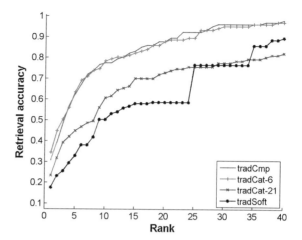

Fig. 8.21 CMC of traditional soft biometrics and when add clothing

at rank 1 is effectively enhanced by all clothing approaches up to 17% in the best case yielded by *tradCat-6*, and the average score up to rank 10 of *tradSoft* is also improved with a significant increase ranging from 9 to 28%. *tradCmp* is the best in terms of the decidability index and achieves the highest overall rank, indicating a better separation between genuine and imposter distributions and more successful discrimination between subjects. *tradCat-6* is the first to achieve 100% at rank 58. Figure 8.22 illustrates that *tradCmp* followed by *tradCat-6* receive the best ROC scores with less errors outperforming *tradSoft* and *tradCat-21*. The performance of *tradCat-21* is low compared with the other clothing approaches, though it attains some recognition capability improving up to rank 34.

Figure 8.23 shows two query examples of subject retrieval using *tradCmp* approach achieving the highest overall retrieval performance. In both examples, the top left corner image represents a side-view image used to derive the side-query descriptions. The remaining numbered front-view images represent the top *k* retrieved subjects from the test gallery, ordered based on their similarity to the query-description, where the query image and the correctly retrieved subject are bordered in yellow. In the first example on the left a query subject was correctly retrieved at rank 1, whereas in the second example on the right another subject was retrieved only at rank 7. In the second (right) example, it can be observed that all retrieved subjects are very similar in their clothing such as sleeve and leg length, and these similarities are correctly reflected by the match. It appears that in such a case, the strong similarities in comparative clothing traits and in the four traditional biometrics, may result in a confusion between such similar subjects. Despite of the desirable objective of retrieving the correct subject as the top match (rank 1), the retrieval of the correct subject within a small list (e.g. 10 subjects) appears reasonably successful and certainly will be useful to narrow the search. So, retrieval may not always answer the question is the top match correct? but instead could answer "is the correct answer in the top *k* matches?".

Fig. 8.22 ROC of
traditional soft biometrics
and when add clothing

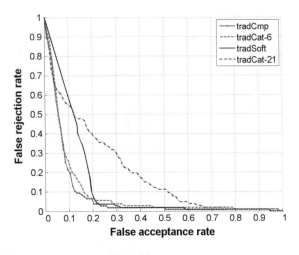

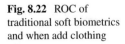

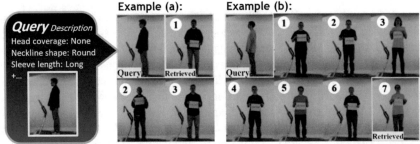

Fig. 8.23 Retrieval query example using *tradCmp*, Example **a** a subject correctly retrieved at rank 1, Example **b** a subject retrieved but at rank 7

8.5.2.2 Fusion of Soft Clothing and Soft Body Biometrics

Clothing descriptions are reused here to supplement the front/side observable 17 soft body descriptions including the four traditional traits (see Table 8.11). Table 8.13 reports metric scores of all approaches, and Fig. 8.24 compares the CMC performance (up to rank 40) of these approaches. The clothing approaches *softCat-6* and *softCmp* respectively provide the highest performance, while *softCat-6* gains the best scores in all evaluation metrics but d' and EER, and outperforms all approaches as also reflected by the ROC in Fig. 8.25. The rank 1 retrieval of *softBody* is obviously enhanced when adding clothing from about 86 to 92% by *softCmp*, and to 94% by *softCat-6*, given in Table 8.13. As such, the clothing analysis can effectively augment soft body descriptions. This enhanced performance is reflected in the class distributions and errors shown in Fig. 8.25 which confirm the potency *softCat-6* and *softCmp* labels. Since Query descriptions have been acquired using side-view images,some items

Fig. 8.24 CMC (rank 40) of soft body biometrics and when add clothing

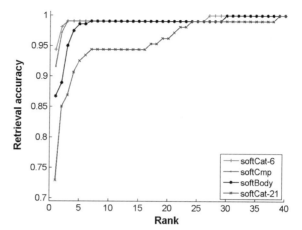

Fig. 8.25 ROC of soft body biometrics and when add clothing

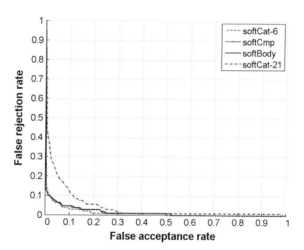

are difficult to observe or occluded such as neckline shape or size and belt presence. As *softCat-21* consists of all clothing traits including those affected traits, this can produce, for affected traits, inconsistent descriptions and undesirable increase in within-class variance. That appears to be a reason for the low performance compared with the other approaches. Another reason could be the noise caused by adding a large number of clothing traits to another large number of body traits.

Table 8.13 Performance metrics of soft body biometrics and when add soft clothing traits

Approach	Top rank	AVG sum match scores up to rank		100% accuracy achieved at rank	EER	AUC	d'	Overall rank
	= 1	= 10	= 128					
softBody	0.868	0.962	0.996	30	**0.064**	0.015	**3.442**	3
softCat-21	0.729	0.899	0.987	39	0.108	0.042	1.802	4
softCat-6	**0.943**	**0.985**	**0.998**	**27**	0.070	**0.012**	3.117	**1**
softCmp	0.916	0.981	0.997	30	0.070	0.014	3.406	2

8.6   Conclusion

This chapter provides an overview of different clothing attribute analyses and focuses on those for biometric purposes. Extended analysis has been conducted with detailed representation of soft clothing biometrics and their capabilities in person identification and retrieval. A methodology framework was offered for soft clothing biometrics derivation and their performance evaluation. Furthermore, clothing feature space was studied via performance analysis and empirical investigation of the capabilities of soft clothing biometrics in subject identification and retrieval, leading to a perceptive guide for feature subset selection and enhanced performance. A number of vignette experiments were conducted for evaluation and comparison of the viability of different categorical and comparative soft clothing approaches.

In this chapter we have shown how semantic clothing attributes can be analysed and indeed be used purely as soft biometrics, resulting in subject identification and retrieval by using them as major biometric traits either in isolation or in fusion. The ensuing experimental results highlighted the potency of soft clothing biometrics in identification and challenging retrieval either when used alone or when used to supplement other biometric traits.

Soft clothing traits can be very useful as a major cue or ancillary information for identity in scenarios suffering from high variability issues. Since clothing traits are more likely visible and perceivable soft biometrics in surveillance videos, they could be invariant to partial occlusion or viewpoint change and effective in challenging cases, such as when criminals cannot be identified by their faces. The capability of verbal identification or retrieval from images and videos can pave the way for a verity of useful applications that can for example search surveillance data of a crime scene to match people to potential suspects described verbally by eyewitnesses, since soft clothing biometrics were meant to bridge the semantic gap between machine and human.

References

1. Adjeroh, D., Deng, C., Piccirilli, M., et al.: Predictability and correlation in human metrology. In: IEEE International Workshop on Information Forensics and Security (WIFS) (2010)
2. Almudhahka, N., Nixon, M., Hare, J.: Human face identification via comparative soft biometrics. In: IEEE International Conference on Identity, Security and Behavior Analysis (ISBA) (2016)
3. Ao, M., Yi, D., Lei, Z., et al.: Face recognition at a distance: system issues. In: Handbook of Remote Biometrics, pp. 155–167. Springer (2009)
4. Arigbabu, O.A., Ahmad, S.M.S., Adnan, W.A.W., et al.: Integration of multiple soft biometrics for human identification. Pattern Recognit. Lett. **68**, 278–287 (2015)
5. Arigbabu, O.A., Ahmad, S.M.S., Adnan, W.A.W., et al.: Recent advances in facial soft biometrics. In: The Visual Computer (2014)
6. Avraham, T., Lindenbaum, M.: Learning appearance transfer for person re-identification. In: Person Re-Identification, pp. 231–246. Springer (2014)

7. Best-Rowden, L., Bisht, S., Klontz, J.C., et al.: Unconstrained face recognition: establishing baseline human performance via crowdsourcing. In: IEEE International Joint Conference on Biometrics (IJCB) (2014)
8. Bialkowski, A., Denman, S., Lucey, P., et al.: A database for person re-identification in multi-camera surveillance networks. In: IEEE International Conference on Digital Image Computing Techniques and Applications (DICTA 12) (2012)
9. Bossard, L., Dantone, M., Leistner, C., et al.: Apparel classification with style. In: Asian Conference on Computer Vision (ACCV) (2013)
10. Cai, S., Wang, F., Quan, L.: How fashion talks: clothing-region-based gender recognition. In: Progress in Pattern Recognition, Image Analysis, Computer Vision, and Applications, pp. 515–523. Springer (2014)
11. Chao, X., Huiskes, M.J., Gritti, T., et al.: A framework for robust feature selection for real-time fashion style recommendation. In: Proceedings 1st International Workshop on Interactive Multimedia for Consumer Electronics, pp. 2037–2041. ACM (2009)
12. Chapman, G.B., Johnson, E.J.: Incorporating the irrelevant: anchors in judgments of belief and value. Heuristics and Biases: The Psychology of Intuitive Judgment. New York, Cambridge University Press (2002)
13. Chen, H., Gallagher, A., Girod, B.: Describing clothing by semantic attributes. In: European Conference on Computer Vision (ECCV), pp. 609–623 (2006)
14. Chen, X., An, L., Bhanu, B.: Soft-biometrics and reference set integrated model for tracking across cameras. In: Distributed Embedded Smart Cameras, pp. 211–230. Springer (2014)
15. Chen, X., Bhanu, B.: Soft biometrics integrated multi-target tracking. In: Proceedings International Conference on Pattern Recognition (ICPR) (2014)
16. Cushen, G.A., Nixon, M.S.: Real-time semantic clothing segmentation. In: Advances in Visual Computing, pp. 272–281. Springer (2012)
17. Dantcheva, A., Dugelay, J-L.: Frontal-to-side face re-identification based on hair, skin and clothes patches. In: 8th IEEE International Conference on Advanced Video and Signal-Based Surveillance (AVSS), pp. 309–313 (2011)
18. Dantcheva, A., Dugelay, J-L., Elia, P.: Person recognition using a bag of facial soft biometrics (BoFSB). In: IEEE International Workshop on Multimedia Signal Processing (MMSP), pp. 511–516 (2010)
19. Dantcheva, A., Elia, P., Ross, A.: What else does your biometric data reveal? a survey on soft biometrics. IEEE Trans. Inf. Forensics Secur. (TIFS) **11**(3), 441–467 (2016)
20. Dantcheva, A., Velardo, C., D'angelo, A., et al.: Bag of soft biometrics for person identification. Multimed. Tools Appl. **51**, 739–777 (2011)
21. Demirkus, M., Garg, K., Guler, S.: Automated person categorization for video surveillance using soft biometrics. In: Proceedings SPIE, pp. 76670P–76671P (2010)
22. Deng, C., Cunjian, C., Piccirilli, M., et al.: Can facial metrology predict gender? In: IEEE International Joint Conference on Biometrics (IJCB) (2011)
23. Deng, Y., Luo, P., Loy, C.C., et al.: Pedestrian attribute recognition at far distance. In: Proceedings of the ACM International Conference on Multimedia, pp. 789–792 (2006)
24. Denman, S., Bialkowski, A., Fookes, C., et al.: Determining operational measures from multi-camera surveillance systems using soft biometrics. In: Proceedings 8th IEEE International Conference on Advanced Video and Signal-Based Surveillance (AVSS), pp. 462–467 (2011)
25. Denman, S., Fookes, C., Bialkowski, A., et al.: Soft-biometrics: unconstrained authentication in a surveillance environment. In: Digital Image Computing: Techniques and Applications (DICTA'09), pp. 196–203 (2009)
26. Denman, S., Halstead, M., Fookes, C., et al.: Searching for people using semantic soft biometric descriptions. Pattern Recognit. Lett. **68**, 306–315 (2015)
27. Doretto, G., Sebastian, T., Tu, P., et al.: Appearance-based person reidentification in camera networks: problem overview and current approaches. J. Ambient Intell. Humaniz. Comput. **2**, 127–151 (2011)
28. Ellis, H.D., Deng, C., Piccirilli, M., et al.: Practical aspects of face memory. In: Eyewitness Testimony: Psychological Perspectives, pp. 12–37. Cambridge University Press (1984)

29. Field, A.: Discovering statistics using SPSS. Sage Publications (2009)
30. Flin, R.H.: Tall stories: Eyewitnesses' ability to estimate height and weight characteristics. Hum. Learn. J. Pract. Res. Appl. (1986)
31. Gallagher, A.C., Chen, T.: Clothing cosegmentation for recognizing people. In: IEEE Conference on Computer Vision and Pattern Recognition (CVPR) (2008)
32. Gray, D., Brennan, S., Tao, S.: Evaluating appearance models for recognition, reacquisition, and tracking. In: IEEE International Workshop on Performance Evaluation for Tracking and Surveillance (2007)
33. Guo, G.: Human age estimation and sex classification. In: Video Analytics for Business Intelligence, pp. 101–131. Springer (2012)
34. Halstead, M., Denman, S., Sridharan, S., et al.: Locating people in video from semantic descriptions: a new database and approach. In: Proceedings International Conference on Pattern Recognition (ICPR), pp. 469–481 (2014)
35. Jaha, E.S., Ghouti, L.: Color face recognition using quaternion PCA. In: 4th International Conference on Imaging for Crime Detection and Prevention (ICDP) (2011)
36. Jaha, E.S., Nixon, M.S.: Analysing Soft Clothing Biometrics for Retrieval. In: International Workshop on Biometrics (BIOMET). Springer (2014)
37. Jaha, E.S., Nixon, M.S.: Soft biometrics for subject identification using clothing attributes. In: IEEE International Joint Conference on Biometrics (IJCB) (2014)
38. Jaha, E.S., Nixon, M.S.: Viewpoint invariant subject retrieval via soft clothing biometrics. In: IEEE International Conference on Biometrics (ICB), pp. 73–78 (2015)
39. Jain, A.K., Dass, S.C., Nandakumar, K.: Can soft biometric traits assist user recognition? In: Proceedings SPIE 5404, Biometric Technology for Human Identification. International Society for Optics and Photonics, pp. 561–572 (2004)
40. Jain, A.K., Nandakumar, K., Lu, X., et al.: Integrating faces, fingerprints, and soft biometric traits for user recognition. In: Biometric Authentication, pp. 259–269. Springer (2004)
41. Jain, A.K., Park, U.: Facial marks: Soft biometric for face recognition. In: 16th IEEE International Conference on Image Processing (ICIP), pp. 37–40 (2009)
42. Klare, B.F., Klum, S., Klontz, J.C., et al.: Suspect identification based on descriptive facial attributes. In: IEEE International Joint Conference on Biometrics (IJCB) (2014)
43. Kovashka, A., Parikh, D., Grauman, K.: Whittlesearch: image search with relative attribute feedback. In: IEEE Conference on Computer Vision and Pattern Recognition (CVPR), pp. 2973–2980 (2012)
44. Layne, R., Hospedales, T.M., Gong, S.: Attributes-Based Re-identification. In: Person Re-Identification, pp. 469–481. Springer (2014)
45. Layne, R., Hospedales, T.M., Gong, S.: Person Re-identification by Attributes. In: Proceedings British Machine Vision Conference (BMVC) (2012)
46. Layne, R., Hospedales, T.M., Gong, S.: Towards person identification and re-identification with attributes. In: European Conference on Computer Vision (ECCV), pp. 402–412, Springer (2012)
47. Li, A., Liu, L., Wang, K., et al.: Clothing attributes assisted person re-identification. IEEE Trans. Circuits Syst. Video Technol. **5**, 869–878 (2014)
48. Li, A., Liu, L., Yan, S.: Person re-identification by attribute-assisted clothes appearanc. In: Person Re-Identification, pp. 119–138. Springer (2014)
49. Lindsay, R.C., Martin, R., Webber, L.: Default values in eyewitness descriptions: a problem for the match-to-description lineup foil selection strategy. Law Hum. Behav. **18**(5), 527 (1994)
50. Liu, S., Feng, J., Song, Z., et al.: Hi, magic closet, tell me what to wear! In: Proceedings ACM International Conference on Multimedia (ACM MM12) (2012)
51. Liu, S., Song, Z., Liu, G., et al.: Street-to-shop: cross-scenario clothing retrieval via parts alignment and auxiliary set. In: EEE Conference on Computer Vision and Pattern Recognition (CVPR) (2012)
52. Macleod, M.D., Frowley, J.N., Shepherd, J.W.: Whole body information: its relevance to eyewitnesses. In: Adult eyewitness testimony. Cambridge University Press (1994)

53. Martinho-Corbishley, D., Nxion, M.S., Carter, J.N.: Soft biometric retrieval to describe and identify surveillance images. In: IEEE International Conference on Identity, Security and Behavior Analysis (ISBA) (2016)
54. Mery, D., Bowyer, K.: Automatic facial attribute analysis via adaptive sparse representation of random patches. Pattern Recognit. Lett. **68**, 260–269 (2015)
55. Nambiar, A., Bernardino, A., Nascimento, J.: Shape Context for soft biometrics in person re-identification and database retrieval. Pattern Recognit. Lett. **68**(12), 297–305 (2015)
56. Niinuma, K., Park, U., Jain, A.K.: Soft biometric traits for continuous user authentication. IEEE Trans. Inf. Forensics Secur. (TFIS) **5**(4), 771–780 (2010)
57. Nixon, M.S., Correia, P., Nasrollahi, K., et al.: On Soft Biometrics. Pattern Recognit. Lett. **68**, 218–230 (2015)
58. O'toole, A.J., Deng, C., Piccirilli, M., et al.: Psychological and neural perspectives on human face recognition. In: Handbook of face recognition, pp. 349–369. Springer (2005)
59. Parikh, D., Grauman, K.: Relative attributes. In: IEEE International Conference on Computer Vision (ICCV) (2011)
60. Park, U., Jain, A.K.: Face matching and retrieval using soft biometrics. In: IEEE Trans. Inf. Forensics Secur. (TFIS) **5**, 406–415 (2010)
61. Poh, N., Wong, R., Kittler, J., et al.: Challenges and research directions for adaptive biometric recognition systems. In: Advances in Biometrics, pp. 753–764. Springer (2009)
62. Pohjalainen, J., Rsnen, O., Kadioglu, S.: Feature selection methods and their combinations in high-dimensional classification of speaker likability. Comput. Speech Lang. **29**, 145–171 (2015)
63. Reid, D.A., Nixon, M.S., Stevenage, S.: Soft biometrics; human identification using comparative descriptions. IEEE Trans. Pattern Anal. Mach. Intell. (TPAMI) **36**(6), 1216–1228 (2014)
64. Reid, D.A., Nixon, M.S.: Human identification using facial comparative descriptions. In: IEEE International Conference on Biometrics (ICB) (2013)
65. Reid, D.A., Nixon, M.S.: Imputing human descriptions in semantic biometrics. In: Proceedings 2nd ACM Workshop on Multimedia in Forensics, Security and Intelligence, pp. 25–30 (2010)
66. Reid, D.A., Nixon, M.S.: Using comparative human descriptions for soft biometrics. In: IEEE International Joint Conference on Biometrics (IJCB) (2011)
67. Reid, D.A., Nixon, M.S., Stevenage, S.: Identifying humans using comparative descriptions. In: Proceedings 4th International Conference on Imaging for Crime Detection and Prevention (ICDP) (2011)
68. Samangooei, S., Guo, B., Nixon, M.S.: The use of semantic human description as a soft biometric. In: Proceedings 2nd IEEE International Conference on Biometrics: Theory, Applications and Systems (BTAS) (2008)
69. Samangooei, S., Nixon, M.S.: On semantic soft-biometric labels. In: Biometric Authentication, pp. 3–15. Springer (2014)
70. Samangooei, S., Nixon, M.S.: Performing content-based retrieval of humans using gait biometrics. Multimed. Tools Appl. **49**, 195–212 (2010)
71. Schumann, A., Monari, E.: A soft-biometrics dataset for person tracking and re-identification. In: 11th IEEE International Conference on Advanced Video and Signal Based Surveillance (AVSS), pp. 193–198 (2014)
72. Shi, Z., Hospedales, T.M., Xiang, T.: Transferring a semantic representation for person re-identification and search. In: IEEE Conference on Computer Vision and Pattern Recognition (CVPR) (2015)
73. Shutler, J., Grant, M., Nixon, M.S.: On a large sequence-based human gait database. In: Proceedings Recent Advances in Soft Computing (RASC02) (2002)
74. Siddiquie, B., Feris, R.S., Davis, L.S.: Image ranking and retrieval based on multi-attribute queries. In: IEEE Conference on Computer Vision and Pattern Recognition (CVPR), pp. 801–808 (2011)
75. Sunderrajan, S., Manjunath, B.: Context-aware hypergraph modeling for re-identification and summarization. IEEE Trans. Multimed. **18**, 51–63 (2016)

76. Thornton, J., Baran-Gale, J., Butler, D., et al.: Person attribute search for large-area video surveillance. In: IEEE International Conference on Technologies for Homeland Security (HST), pp. 55–61
77. Toews, M., Arbel, T.: Detection, localization, and sex classification of faces from arbitrary viewpoints and under occlusion. IEEE Trans. Pattern Anal. Mach. Intell. (TPAMI) **31**, 1567–1581 (2009)
78. Tome, P., Fierrez, J., Vera-Rodriguez, R., et al.: Soft biometrics and their application in person recognition at a distance. IEEE Trans. Inf. Forensics Secur. (TFIS) **9**(3), 464–475 (2014)
79. Vaquero, D.A., Feris, R.S., Tran, D., et al.: Attribute-based people search in surveillance environments. In: IEEE Workshop on Applications of Computer Vision (WACV) (2009)
80. Vezzani, R., Baltieri, D., Cucchiara, R.: People reidentification in surveillance and forensics: a survey. ACM Comput. Surv. (CSUR) **46**(2) (2013)
81. Vrij, A., Pannell, H., Ost, J.: The influence of social pressure and black clothing on crime judgements. Psychol. Crime Law **11**(3), 265–274 (2005)
82. Wang, H., Bao, X., Choudhury, R.R., et al.: InSight: recognizing humans without face recognition. In: Proceedings 14th Workshop on Mobile Computing Systems and Applications (2013)
83. Wang, T.Y., Kumar, A.: Recognizing human faces under disguise and makeup. In: IEEE International Conference on Identity, Security and Behavior Analysis (ISBA) (2016)
84. Wang, Y.-F., Chang, E.Y., Cheng, K.P.: A video analysis framework for soft biometry security surveillance. In: Proceedings 3rd ACM International Workshop on Video Surveillance & Sensor Networks, pp. 71–78 (2005)
85. Yang, M., Yu, K.: Real-time clothing recognition in surveillance videos. In: Proceedings 18th IEEE International Conference on Image Processing (ICIP), pp. 2937–2940 (2004)
86. Zhang, H., Beveridge, J.R., Draper, B.A., et al.: On the effectiveness of soft biometrics for increasing face verification rates. Comput. Vis. Image Underst. **137**, 50–62 (2015)
87. Zhang, L., Kalashnikov, D.V., Mehrotra, S.: Context assisted person identification for images and videos. Handbook Pattern Recognit. Comput. Vis. (2015)
88. Zhang, L., Kalashnikov, D.V., Mehrotra, S., et al.: Context-based person identification framework for smart video surveillance. Mach. Vis. Appl. **25**(7), 1711–1725 (2014)
89. Zhang, W., Begole, B., Chu, M., et al.: Real-time clothes comparison based on multi-view vision. In: Proceedings 2nd ACM/IEEE International Conference on Distributed Smart Cameras (ICDSC) (2008)
90. Zhou, X., Bhanu, B., Han, J.: Human recognition at a distance in video by integrating face profile and gait. In: Face Biometrics for Personal Identification, pp. 165–181 (2007)
91. Zhu, J., Liao, S., Lei, Z., et al.: Multi-label convolutional neural network based pedestrian attribute classification. Image Vis. Comput. (2016)

Chapter 9
Performance Evaluation of Video Analytics for Traffic Incident Detection and Vehicle Counts Collection

Kitae Kim, Slobodan Gutesa, Branislav Dimitrijevic, Joyoung Lee, Lazar Spasovic, Wasif Mirza and Jeevanjot Singh

Abstract Current incident detection and traffic monitoring method using closed-circuit television (CCTV) cameras meets with limitations as the coverage of CCTV cameras rapidly expands. In general, traffic operators at Traffic Operation Center (TOC) have to manage and monitor numerous CCTV cameras deployed on roadways. Thus, many transportation agencies consider the use of video analytics system to reduce incident detection time and minimize traffic impacts, but they also want to validate the performance of the video analytics system whether it can work with their existing video surveillance infrastructure before procuring the system. To that end, a pilot study was designed and conducted to evaluate the accuracy of a video analytics product by integrating with CCTV cameras deployed on highways. The pilot study was designed to evaluate the accuracy of video analytics in detecting incidents and collecting traffic counts. The test results show that the performance of video analytics is significantly impacted by video quality and other environmental factors such as lighting and weather conditions.

9.1 Introduction

Rapid verification of incidents on roadways is a critical factor to dispatch incident response teams to reduce roadway congestions. According to a survey that questioned the extent of use of technologies and methods for incident verification, CCTV cameras, cellular phones, and highway patrol communication are the respective top three the state-of-the-practice methods [1]. Specifically, the traffic operators still rely heavily on CCTV cameras to visually verify and assess congestion and the incident scenes.

As the coverage of CCTV cameras on roadways is being fast expanded, operators are not capable of monitoring many camera images simultaneously. Thus, the use

K. Kim (✉) · S. Gutesa · B. Dimitrijevic · J. Lee · L. Spasovic
New Jersey Institute of Technology, Newark, NJ 07102, USA
e-mail: kitae.kim@njit.edu

W. Mirza · J. Singh
New Jersey Department of Transportation, Ewing Township, NJ 08618, USA

© Springer International Publishing AG 2017
C. Liu (ed.), *Recent Advances in Intelligent Image Search and Video Retrieval*,
Intelligent Systems Reference Library 121, DOI 10.1007/978-3-319-52081-0_9

of the video analytics (VA) system has attracted considerable attention for traffic monitoring and incident management [2]. Vision-based technology is a state-of-the-art approach with advantages of easy maintenance, real-time visualization, and high flexibility compared to other existent technologies. Previously, a number of commercial VA systems have been introduced to traffic monitoring and incident detection on freeways, signalized intersections, bridges, and tunnels [3–6].

VA is one of the non-intrusive technologies for large-scale implementation of traffic monitoring and control and incident management strategies. VA processes and analyzes videos to identify various incident events, including stopped-vehicle, wrong-way vehicle, congestions, pedestrian on roadway, or debris in real-time. VA can also analyze videos to extract various traffic data information, including volume, speed, lane occupancy, and vehicle classification [7].

Results of the literature search conducted revealed that the majority of previous research focused on the video system architecture, detection algorithm, motion pattern recognition, behavior, event analysis, etc. Limited information is available on the performance variation of the VA system integrated with various quality videos [8]. Currently digital and IP based cameras are widely applied in the transportation industry due to picture quality, increased scalability, and more flexible infrastructure. However, many analog cameras that were deployed previously are still prevalent. In addition to video quality, setting the appropriate compression method is also a significant factor to achieve satisfactory quality for the given bandwidth. Although VA requires reliable transmission of high quality video over networks, videos from the cameras are required to be encoded to comparatively low quality not to exceed the size of the given bandwidth [9].

Clear visibility of objects of interest in videos is a fundamental condition for a successful video-based detection [10]. Variation in illumination of the videos may deteriorate the effectiveness of VA. It is widely known that sight disturbing lighting conditions, such as sun glare, shadows, and darkness are easily observed in the video feeds based on the time of day and the direction of the cameras. Additionally, adverse weather conditions, including heavy rain and snow (e.g., raindrop and ice on camera housing) are also common drawbacks to trigger false incident alarms in incident detection.

To address the issues aforementioned, this study investigates the performance of VA in detecting incidents and collecting data considering the variation of video quality, lighting and weather conditions. To achieve this objective, a pilot study was established using a commercial VA product and the existing CCTV cameras in operation which are deployed on highways.

9.2 Relevant Studies

The use of VA for automatic incident detection has attracted much attention in the past two decades, and many previous studies assessed the performance of many commercial systems. However, limited previous studies evaluated the impact of factors on system accuracy and reliability. Most of the evaluation studies focused on the accuracy of measured traffic data such as volume, speed, and occupancy [8]. Sources of errors, including camera location, field of camera view, time of day, and environmental conditions, were investigated and analyzed [6–8]. It was found that the camera height is the most significant factor that affects the performance of VA because low camera height may cause one vehicle to obscure the view of other vehicles. In the meantime, a couple of studies presented techniques to enhance video quality for better performance [11] developed a fully digital auto-focusing (FDAF) system associated with digital image stabilization for video quality improvement. Sharrab and Sarhan [12] discussed that video denoising techniques can be adopted into VA. Murino et al. [13] presented an automatic intelligent approach that enables a camera to change its parameters automatically to acquire better image.

Parkany and Xie [14] investigated the impact of the camera height on the accuracy of vehicle detection. In this study, a height of 30–40 ft was considered as the optimal height for vehicle detection. It was found that while increased height minimizes the effects of occlusion and improves the cameras view, excessive sway may occur during windy conditions if the camera is installed too high.

Wan et al. [15] tested a commercial video analytics product, Abacus from Iteris, in various weather and lighting conditions. In this study, it was discussed that the height of camera should be sufficient and the horizontal distance from the camera to the detection zone must not be excessive so that horizontal occlusion is minimized. The study demonstrated that camera placement (e.g., perspective) is crucial to the performance of the VA system. The camera height of 12.2–18.2 m (40–60 ft) is ideal for accurate performance. In addition, the quality of input video was identified as an important factor for successful vehicle detection.

Grant et al. [16] identified the factors adversely impacting the accuracy of traffic data collected by VA. They estimated the accuracy of volume, classification, and speeds with manual counts. It was found that the geometric layout of the detection area in the video image was the most significant factor that affected the accuracy of volume data. It was also discussed that the slanted camera view could be a reason for triggering false volume counts because tall heavy vehicles can be detected by adjacent lanes. Additionally, it was also found that headlight beams can trigger false counts at night. Finally, environmental factors such as rain, wind, and snow were identified as reasons to degrade the accuracy of the traffic data measurement.

Prevedouros et al. [8] discussed that possible factors affecting the accuracy of most video detection systems. During the nighttime period, headlight beams, halos, and other lighting effects significantly impacted the accuracy of video detection. In addition, the impact of sun glare and shadows on the accuracy of video detections may considerably trigger false calls, which are critical in the performance of incident detection.

A research study conducted by the Minnesota Department of Transportation (MNDOT) evaluated the accuracy of traffic data measured by the two commercial VA systems, Traficon and Autoscope [17]. These systems were tested at several locations with different installation settings to assess the incident detection performance with various camera heights. It was found that the accuracy of incident detection was different across lanes because the surveillance cameras were generally located in the shoulder area of highways. It was also found that the performance of the Traficon system was better during off-peak than peak periods; it detected volumes with an error between 5 and 10%. The detection error in speed was 8–12%, while the Autoscope volume detection error was between 3 and 10% for most heights and was lower than 6% for speed in most cases.

The importance of video quality on the system performance was examined by Preisen and Deeter [18]. A research study funded by the Enterprise Transportation Pooled Fund program investigated the accuracy of VA systems in detecting incidents and collecting traffic counts. It was discussed that if an input video is choppy, it will not be processed by VA as accurately as a feed that is stream with minimal interruption. The study also discussed the importance of the communication methods used to deliver video streams from cameras in the field to VA servers being reliable and stable. The authors noted that the importance of the communication method should not be underestimated for the success of VA performance.

Surveillance cameras in poor circumstances such as narrow communication bandwidth and old CCTV cameras can only provide poor quality video feeds to VA. Some VA systems are equipped with a self-awareness function which diagnoses video quality and alarms for system maintenance with poor video quality [11].

Chintalacheruvu and Muthukumar [2] examined the effect of fluctuating video brightness on the efficiency of the VA system. They proposed a vehicle detection system for ITS application based on the Harris-Stephen Corner Method (HSCM) that requires fewer calibrations and sensitiveness to illumination changes. The developed system was tested, and the performance results were analyzed using a set of video feeds at 1 min intervals. The video feeds varied in brightness (e.g., captured in different time periods of the day), camera mount height, camera view angle, and region of view. The developed system was implemented on an embedded computer platform. The study results found that the developed system can determine vehicle counts and speeds from low resolution video feeds in real-time under various brightness conditions with very little configuration and calibration requirements.

Martin et al. [19] investigated the effect of increment weather on the accuracy of vehicle detection. It was found that dark vehicles are not often detected under increment weather conditions such as rain or snow. Because the dark vehicles blend in with the wet asphalt, the detection algorithm cannot recognize the difference between them. It was also found that shadows from trees, buildings, power cables and other vehicles were sometimes detected as a vehicle, resulting in false alarms. Moreover, it was discussed that vehicle headlights tend to activate the detectors, triggering a false call especially during the night and in bad weather conditions. In addition, glare problems could cause the iris of the lens to shut, bringing down the detection accuracy of the system.

9.3 Design of Pilot Test

An evaluative pilot study was designed to assess the overall VA system effectiveness in detecting traffic incidents and collecting vehicle counts data under various environments, including different video quality, weather, and lighting conditions. This section describes the pilot study design such as system integration, test bed selection, camera view presets, incident detection rule setting, system validation, and performance analysis. Figure 9.1 shows the process of the pilot study and the parameters determined for each step. The framework developed for this pilot study is comprised of three major components: system setup, validation, and analysis.

9.3.1 System Integration

The initial system setup was to integrate the VA server with the video streams that were provided from the primary video streaming server. However, a critical issue in integration of the VA system into the primary streaming server was found during the system installation. The video streaming protocol required by the studied VA system is either the Real Time Streaming Protocol (RTSP) or the Real-time Transport

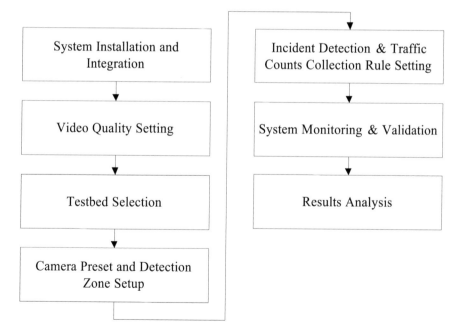

Fig. 9.1 Process of pilot test

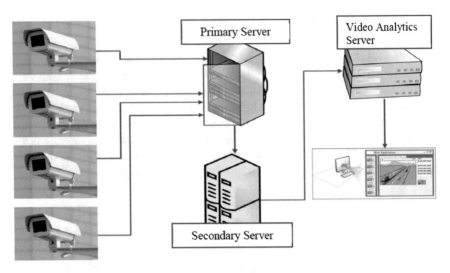

Fig. 9.2 Configuration of video analytics setup

Protocol (RTP), but the available video protocol from the primary server is the Real-time Transport Protocol (RTP).

Due to this reason, the VA system was integrated with the secondary streaming server (see Fig. 9.2), which is connected into the traffic surveillance network to provide video streams originating from CCTV cameras to the video wall in the traffic operation centers. The secondary streaming server supports the option to both push and pull the Real Time Message Protocol (RTMP) streams, as well as supporting the Real Time Streaming Protocol (RTSP) and the Real-time Transport Protocol (RTP), depending on the usage. In addition, the inbound access can be enabled to identify individuals or ranges of IP addresses via the web interface. As a result of this flexibility, the VA server was integrated with the video streams from the video streaming server for the pilot study.

9.3.2 Video Quality Setting

The quality of videos currently available from the secondary video streaming server were much lower than the video quality requirements of the studied VA system. The video streaming server used in this study is receiving and disseminating low quality videos due to the bandwidth capacity of the networks and the server. It is desirable to test the VA system using high quality videos, but it is also required to test the VA system with the existing video quality to figure out whether the VA system accurately performs with the existing video quality because the cost of changing the network setting is expensive. Thus, this pilot study was designed to examine the VA system performance with three different video qualities.

Table 9.1 Video quality setup per setting

	Setting A	Setting B	Setting C
Resolution (pixel)	320×258	640×482	352×482
Frame rate (fps)	15	15	30
Codec	H. 264	H. 264	H. 264
Bitrate (kbps)	<100	<900	>1,000

The three video quality settings were established in the design process, which includes: (1) low quality - available from the secondary video streaming server, (2) medium quality - temporarily increased video quality from the secondary server, and (3) high quality - ideal video quality for the tested VA system. The pilot study includes the experiments that were performed with each of these video quality settings, and each experiment assessed the performance of the VA system considering weather and lighting conditions.

As summarized in Table 9.1, it was defined that the low quality video is Setting A, the medium quality video is Setting B, and the high quality video is Setting C. Video resolution, frame rate, and bit-rate are different by each setting. The quality of Setting B was enhanced by increasing the resolution size. It is worth noting that the increased quality (i.e., resolution) would cause the communication network to overload and impact the outbound bandwidth of the secondary streaming server, which in turn will adversely impact the server stability.

Besides resolution, a video's appearance depends on its frame rate and bit-rate. Frame rate indicates the number of still images over the span of a second, which determines the smoothness of the video. Bit-rate serves an indicator of overall video quality, which is increased with higher resolutions and frame rates. Bit-rate also varies by the strength of compression. A heavily compressed video contains a lower bit-rate than a lightly compressed video. In addition, a higher frame rate also increases the bit-rate because more data is included in the additional frames [20]. Based on this, the quality of Setting C was determined by increasing frame rate and bit-rate.

9.3.3 Test Bed Locations

The primary work in this section was to review all of the candidate CCTV cameras. The reviewed camera features include camera type, location (e.g., side shoulder vs. median), height, roadway curvature (e.g., curve vs. straight), etc.

In order to obtain a sufficient incident event sample for the pilot study, two year incident records that occurred within 1,000 ft of each CCTV camera were examined and ranked based on the incident frequency rates. Additionally, traffic conditions operating around cameras during peak and off-peak periods were investigated with historical probe vehicle speed data.

Out of a total of 315 cameras that were reviewed, 21 cameras were selected. The cameras are located in either shoulder or median, and the heights of cameras are 40, 55, or 75 ft. All test bed cameras along with major highways and no cameras at signalized intersections were included because of the scope of the VA systems application.

Additionally, the 16 test bed cameras studied with the Setting A quality are not capable of providing increased quality video due to hardware limitations. Considering all these aspects, four test bed cameras were selected and integrated with Setting B. Lastly, the test with Setting C was performed with a single camera. The videos were recorded at the site and directly provided into the VA server using a VLC player to exclude any impacts from the communication network.

9.3.4 Camera Preset and Detection Zone Set Up

For more accurate video analytics performance, camera presets at a desirable view were established. The recommended ideal camera preset considers the following criteria:

- Keep as little of the sky in view as possible
- Avoid signs blocking the field of view
- Maintain appropriate zoom levels (i.e., vehicles should not appear too big or small)

After integrating the system, camera presets (e.g., calibrated camera positions) were made at a desirable view to ensure that the VA system accurately detects incidents and collects traffic data. One or two presets were made for each of the test bed cameras. Once the camera preset is complete, calibration work was also conducted to define the areas where incident events and vehicles are detected. The detection areas in a cameras video are defined by an outline that is superimposed over the CCTV cameras video image. Two different virtual boundaries were created for incident detection and traffic data collection. As illustrated in Fig. 9.3, the green colored virtual boundaries are for incident detection. Any incident event that occurs within this boundary will be automatically detected. The black colored boundaries are made for each lane to help collect traffic counts and estimate speed.

9.3.5 Incident Detection Rule Setting

The tested VA system allows users to develop various incident detection rules for each camera. Considering the incident type and traffic speed, users can select five different types of incident events to be detected, including:

- Wrong-way vehicles;
- Stopped vehicle;

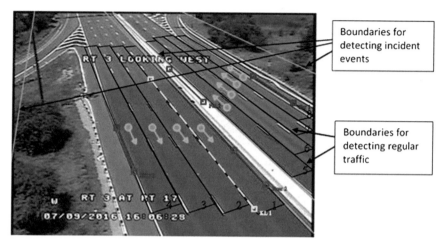

Boundaries for detecting incident events

Boundaries for detecting regular traffic

Fig. 9.3 Virtual detection zones for incident detection and data collection

- Congestion;
- Slow speeds and,
- Pedestrian

For stopped vehicle incident detection, users select a minimum time interval. For example, if a user selects one minute as the stopped vehicle detection interval, vehicles in standstill for less than one minute would not be defined as an incident by the VA system and would not trigger an alert to traffic operators. Similarly, congestion incidents are defined by minimum lane occupancy and slow speed incidents are defined by both maximum speed and minimum duration.

9.3.6 System Validation Approach

An incident validation process was developed to verify positive incident detections, as well as to identify false positive detections and undetected incidents. To ensure an accurate validation, video feeds from the test bed cameras were recorded during the entire pilot study period, and all types of incidents defined in the previous section were manually checked. The video recording was conducted to record CCTV videos every 30 min for 24 h.

The video review results were compared with incident log files created by the VA system. Since it was observed that the test bed cameras frequently strayed from the preset position for traffic operation purposes, the time intervals when cameras were out of the preset positions were excluded from the validation.

The accuracy of traffic volume data was validated separately. Hourly traffic volumes under various weather and lighting conditions were validated by comparing

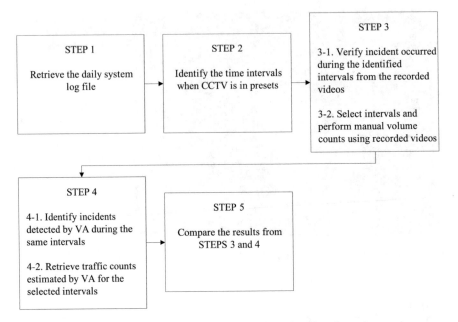

Fig. 9.4 Process of validating detected incidents and estimated traffic volume data accuracy

them with manual counts. The process of validating incident detection and traffic counts is illustrated in Fig. 9.4.

Two measures of effectiveness were used for evaluating the accuracy of incident detection, which are detection rate (DR) and false alarm rate (FAR). DR is the ratio of detected incidents to the total incidents occurred.

$$DR = \frac{number\ of\ detected\ incidents}{number\ of\ actual\ incidents} \times 100 \qquad (9.1)$$

FAR is the percentage of false alarms relative to the total number of alarms triggered.

$$FAR = \frac{number\ of\ false\ alarms}{number\ of\ total\ triggered\ alarms} \times 100 \qquad (9.2)$$

Volume data quality was studied using two statistical measures, mean absolute error (MAE) and mean absolute percent error (MAPE), for 15 min interval.

$$MAE_i = \left| \overline{v_i} - \overline{V_i} \right| \qquad (9.3)$$

where MAE_i is the mean absolute error for the ith 15-min period. $\overline{v_i}$ is the mean volume estimated by VA for the ith 15-min period. $\overline{V_i}$ is the mean ground truth volume for the ith 15-min period.

$$MAPE_i = \frac{1}{n} \sum_{i=1}^{n} \frac{MAE_i}{\overline{V_i}} \qquad (9.4)$$

where n represents the total number of 15-min interval.

9.4   Validation Results

This section describes the analysis of the validation process for incident detection and traffic volume count data collection. The findings of this effort are expected to provide a substantial contribution to the understanding of the value of using video analytics.

9.4.1   Incident Detection

9.4.1.1   Impact of Video Quality

The accuracy of incident detection was measured by DR and FAR considering video quality, weather and lighting conditions. Figure 9.5 shows the results of incident detection associated with three different quality settings. It was found that as the quality of video increases, DR increases, and FAR decreases. During the study with the existing quality videos (i.e., Setting A), a total of 152 incident events occurred, and only 38 incidents of them were detected by the VA system. After increasing the quality of video (i.e., Setting B), DR was considerably improved up to 70.3% while FAR decreased to 12.5%. For the duration of the test with Setting B, a total of 259 incidents were identified from the videos, and 182 incident of them were detected. For the experiment with the ideal quality video (i.e., Setting C), there were 51 incidents, and 41 incidents of them were detected by the VA system, and only 3 false alarms were triggered for this period. It was confirmed that the quality of input video is a very significant factor that determines the VA performance.

9.4.1.2   Impact of Weather Conditions

For the period of the pilot study associated with the Setting A quality videos, there occurred 8 incidents under rain conditions and 2 incidents under fog conditions, but none of these incidents were detected by the VA system. It is not clear whether the undetected incidents were solely impacted by the weather conditions or partially affected by the video quality. On the other hand, during the period of the study with the Setting B quality, a total of 84 incidents occurred under snow conditions, and 58 incidents of the 84 incidents were detected while 3 false alarms were also triggered.

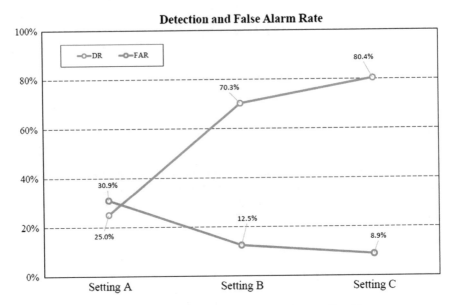

Fig. 9.5 Performance of incident detection associated with various quality videos

It is worth noting that no incident occurred under adverse weather (e.g., rain) condition during the test with the Setting C quality because the test duration was comparatively shorter than the others two setting studies.

9.4.1.3   Impact of Lighting Conditions

The accuracy of incident detection under sun glare and darkness conditions was also examined. As summarized in Table 9.2, a total of 69 and 22 incidents occurred under sun glare and darkness conditions, respectively, for the period of the study with the

Table 9.2 Validation results of incident detection under sun glare and darkness conditions

			Video quality		
			Setting A	Setting B	Setting C
Lighting conditions	Sun glare	Actual incident	69	108	No data available
		Detected incident	2	73	
		False alarm	0	12	
	Darkness	Actual incident	22	66	14
		Detected incident	0	51	11
		False alarm	0	11	0

Setting 1 quality. However, only 2 incidents under sun glare were detected and none of the incidents under darkness conditions was detected. The validation results with the Setting 2 quality shows that 73 out of 108 incidents and 51 out of 66 incidents were successfully detected under sun glare and darkness conditions, respectively. At the same time, however, there were 12 and 11 false alarms triggered under sun glare and darkness conditions, respectively. There occurred 14 incidents under darkness situation during the period of the study with the Setting 3, and 11 incidents of them were detected. Please note that no incidents were detected under sun glare condition during this period.

9.4.1.4 Impact of Camera Height and Position

The impact of camera height and position on the accuracy of incident detection and false positive alarms was also investigated. Four camera heights: 20, 40, 55, and 75 ft, and two positions: shoulder and median, were considered.

As shown in Table 9.3, there was no significant relationship between incident detection and camera height, but the DR at 55 ft was relatively higher than at the other camera heights. Meanwhile, the FAR at 20 ft was 50% although it is not a statistically sound sample size. We also reviewed video recordings to identify whether the undetected incidents or false alarms were triggered by occluded camera views that occurred due to tall vehicles. However, none of the undetected incidents (or false positive alarms) was related with occlusion in the cameras view. In addition, it was not observed whether the camera position: median vs. shoulder, actually affected the degree of incident detection accuracy.

Overall, it was found that while cameras are at or above the recommended camera height, occlusion of the cameras field of view seldom occurred. The minimum camera height recommended by the VA is 40 ft.

9.4.1.5 Discussion of Undetected Incidents

This section analyzes the undetected incidents associated with the ideal video quality that occurred. For the test with Setting C, a total of ten stopped vehicle incidents were not detected, but it is not yet clear exactly what caused these unsuccessful detections. As revealed in the previous sections, the accuracy of the video analytics system is highly impacted by the video quality. Lighting and weather conditions can also impact and deteriorate the quality of the video. Thus, the undetected incidents were investigated by considering various potential conditions that would likely impact the accuracy, which included vehicle pixel size, time of day (e.g., lighting condition), location of vehicles in the image, and the color of the vehicles.

First, the pixel size of all undetected incidents was measured, and it was found that the size of each undetected incident was greater than 10×10 pixels, which is the minimum size requirement to be detected by the VA. Four undetected incidents occurred during the night time period and the other six undetected incidents occurred

Table 9.3 Incident detection results with camera heights and positions

			Actual incident	Detected incident	False alarm	DR (%)	FAR (%)
Setting A	Camera height (ft)	40	6	2	2	33.3	50.0
		55	234	53	11	22.6	17.2
		75	326	99	21	30.0	17.6
	Camera position	Median shoulder	129	27	5	20.0	16.1
			547	174	39	31.8	18.3
Setting B	Camera height (ft)	40	81	61	4	75.3	6.2
		55	160	105	5	65.6	4.5
		75	18	16	17	88.9	51.5
	Camera position	Median shoulder	18	16	17	88.9	51.1
			241	166	9	68.9	5.1
Setting C	Camera height (ft)	40	No data available				
		55	51	41	3	80.4	6.8
		75	No data available				
	Camera position	Median shoulder	No data available				
			51	41	3	80.4	6.8

during the day time. Seven undetected incidents occurred at the gore area as shown in Fig. 9.6. The most likely reason for missed detections in the gore area is that the white striping deteriorated the appearance of vehicles in the image, especially when dealing with light-colored vehicles that stopped at that spot. During night-time periods, the video turns black and white due to the lack of light. In addition, the headlights of the vehicles blurred the image, which produced a poor quality video, eventually resulting in a lower accuracy rate of incident detection and data collection.

9.4.1.6 False Alarms

This section summarizes the observed false alarms and the reasons that triggered the false alarms. A number of false positive alarms were observed over the entire pilot study period regardless of video quality. These false alarms were triggered by not only various lighting conditions such as sun glare, darkness, and shadow, but they were also triggered by rain drops on the CCTV camera enclosure and by accumulated snow on the roads.

Previously it was discussed that as the quality of input increases, the number of false alarms decreases. In addition to the effect of video quality on the false alarms, it was also revealed that false alarms are easily triggered by sun glare, shadow, rain drops, snow flake, and blurry lights at night. In particular, many false alarms of

Fig. 9.6 Examples of undetected incidents occurred at the gore area

slow speed and congestion were triggered when the cameras were out of the preset positions. Thus, cameras are must be returned to the preset positions to reduce false alarms. Examples of these false detection are illustrated in Fig. 9.7.

9.4.2 Traffic Count Data

The accuracy of the traffic count data collected by the VA system was investigated by considering both weather and lighting conditions. The evaluation was conducted with the traffic counts extracted from the videos of Settings B and C. The traffic counts collected with the Setting A video was excluded because the quality of Setting A is not sufficient to extract traffic information. Traffic count validation was conducted for both directions, near side and far side. Please note that the near side represents the roadway closer to the camera position, and the far side indicates the roadway on the opposite side.

First, the accuracy of traffic counts collected from the Setting B video was examined. The accuracy under normal weather conditions was validated. In Table 9.4, the hourly MAE for the near side and the far side were 72 vehicles and 81 vehicles, respectively. The corresponding MAPE were 12.0 and 13.7%, respectively.

(a) Sun glare   (b) Rain drops on the lens surface

(c) Snow   (d) Shadow

(e) Unknown   (f) Blurry headlight

Fig. 9.7 Examples of false positive detection

The overall MAE and MAPE results under rain and heavy snow conditions are poorer than that observed under normal condition. It was found that rain and snow conditions clearly impact the accuracy of traffic counts.

The effect of darkness lighting condition on the traffic counts accuracy was also evaluated. The error between the ground truth and the VA system is 716 vehicles and

Table 9.4 Traffic counts validation results under Normal, Rain, Snow, and Darkness conditions

		Ground truth (veh)		VA (veh)		MAE (veh)		MAPE (%)	
		Near side	Far side	Near side	Far side	Near side	Far side	Near side	Far side
Setting B	Normal	2,335	2,344	2,278	2,188	72	81	12.0	13.7
	Rain	865	428	425	186	110	61	51.1	56.3
	Snow	969	913	1,572	1,588	151	169	95.2	118.4
	Darkness	1,708	3,007	1,307	2,299	100	177	23.5	23.6
Setting C	Normal	3,555	2,618	3,834	2,341	279	313	7.8	12.0
	Rain	3,302	2,839	3,288	2,383	114	456	3.5	16.1
	Darkness	2,645	807	2,057	515	716	292	27.1	36.2

292 vehicles for the near and far side, respectively, and the overall MAPE was 27.1 and 36.2% for near and far side, respectively.

Next, the accuracy of traffic counts collected with the Setting C video was also investigated for normal, rain, and darkness conditions. As shown in Table 9.4, it was found that the overall data accuracy was significantly increased when compared to the results obtained with Setting B. Particularly, the overall MAPE under normal condition was 7.8 and 12.0% for the near and far sides, respectively. It was also revealed that the data accuracy on the near side is much higher than the far side for all conditions. Moreover, the results under rain conditions showed a higher accuracy than those under darkness conditions.

9.5 Conclusions

This study evaluated the performance of video analytics system in detecting incidents and collecting traffic data by considering video quality, weather, and lighting conditions. The study was conducted with the selected 21 test bed cameras on highways, and the input videos were provided from the video streaming server.

The accuracy of the results from the incident detection function confirms that the precision of the system is highly impacted by the quality of the input videos. With the low quality of the videos, the averaged incident detection rate was less than 30%.

However, the accuracy rate of the detected incidents was dramatically increased to 70.3% with the use of the enhanced quality videos (e.g., increased resolution and bitrate); albeit, a number of false positive detections were also triggered. With the ideal quality videos, the detection accuracy was increased to 80.4%. The more notable findings of the study are summarized as follows:

The bitrate of input videos significantly impacts the accuracy of incident detection and traffic data collection. With the existing low bitrate (e.g., less than 300 kbps) videos, the averaged overall detection accuracy rate was 13.8%. However, the averaged overall detection accuracy rate was increased to 70.3% with the videos containing the bitrate ranged between 350 and 910 kbps. When the bitrate was increased over 1,000 kbps, the accuracy was above 80.4%. It was found that a number of false alarms were frequently activated due to reflected sun light, vehicle headlights, rain drops, snowflakes, etc., all of which greatly impact the video image quality.

The overall detection accuracy rates that were associated with both the enhanced quality and the ideal quality videos under darkness conditions was 77.3 and 78.6%, respectively. It should be noted that none of the incidents that occurred under darkness conditions associated with the existing video quality was detected. The findings clearly revealed that the video quality and the lighting conditions are the most critical factors affecting the performance of video analytics.

Several incidents were not detected which happened at the gore zone area. It is deemed that the white stripes in the gore zone degrade the clearance of light colored vehicle images.

The quality of the video was found to be the most significant factor in determining the accuracy of traffic data in this study. With the high quality videos, the MAPE of traffic counts on the near and far sides was 7.8 and 12.0%, respectively, under normal conditions. The MAPE of traffic counts under rain condition was 3.5 and 16.1% for the near and far sides, and the MAPE of traffic counts on the near and far sides under darkness condition was 27.1 and 36.2%, respectively.

- The effect of the camera height on the accuracy of the VA performance is insignificant as long as the camera is positioned at or above the recommended height.
- It is important to note that a single camera deployed with Video Analytics may only be well-suited to one application or use. Depending on the geometric layout, zoom level, and camera angle, a single camera may not support all uses and needs. For example, a camera set to an optimal zoom level for detecting debris or pedestrians may not be able to detect traffic incidents and/or collect traffic data from the roadway lanes.
- For more accurate traffic data collection, it is worth considering deploying Video Analytics with fixed cameras rather than using PTZ cameras that are frequently deviated from their home position for traffic operations.

For the video analytics system to be successful in detecting incidents and collecting traffic counts, it is essential that high quality video be provided during operations. Thus, it is mandatory that the current video streaming structure be improved. As an immediate solution for this problem, it is required that the video quality settings of each CCTV camera be identified and the then settings must be adjusted to increase quality. Additionally, integration of the video analytics system with the primary video streaming server should be established by first resolving the video protocol issue. During the pilot study period, it was often observed that the test bed cameras were deviated from their preset position, which deteriorated the accuracy of the system. Thus, an operations regulation needs to be established to ensure that all cameras return to their preset position after each usage.

The CCTV camera locations would be selected to ensure that a wide variety of roadway types; camera heights and locations; lighting; etc. would be considered in the study.

References

1. Chowdhury, M.A.: Benefit cost analysis of accelerated incident clearance. Cell **136**(2), 215–233 (2007)
2. Chintalacheruvu, N.: Video based vehicle detection and its application in intelligent transportation systems. J. Transp. Technol. **2**(04), 305 (2012)
3. Michalopoulos, P.G.: Vehicle detection video through image processing: the autoscope system. IEEE Trans. Veh. Technol. **40**(1), 21–29 (1991)
4. Gupte, S., Masoud, O., Martin, R., Papanikolopoulos, N.: Detection and classification of vehicles. IEEE Trans. Intell. Transp. Syst. **3**(1), 37–47 (2002)
5. Kanhere, N., Birchfield, S., Sarasua, W., Whitney, T.: Real-time detection and tracking of vehicle base fronts for measuring traffic counts and speeds on highways. Transp. Res. Rec. J. Transp. Res. Board **1993**, 155–164 (2007)
6. Zhang, G., Avery, R., Wang, Y.: Video-based vehicle detection and classification system for real-time traffic data collection using uncalibrated video cameras. Transp. Res. Rec. J. Transp. Res. Board **1993**, 138–147 (2007)
7. Malinovskiy, Y., Wu, Y., Wang, Y.: Video-based vehicle detection and tracking using spatiotemporal maps. Transp. Res. Rec. J. Transp. Res. Board **2121**, 81–89 (2009)
8. Prevedouros, P., Ji, X., Papandreou, K., Kopelias, P., Vegiri, V.: Video incident detection tests in freeway tunnels. Transp. Res. Rec. J. Transp. Res. Board **1959**, 130–139 (2006)
9. Huang, Y., Buckles, B.: Low cost wireless network camera sensors for traffic monitoring. Cell **136**(2), 215–233 (2012)
10. Dong, N., Jia, Z., Shao, J., Li, Z., Liu, F., Zhao, J., Peng, P.: Adaptive object detection and visibility improvement in foggy image. J. Multimedia **6**(1), 14–21 (2011)
11. Jeon, J., Yoon, I., Kim, D., Lee, J., Paik, J.: Fully digital auto-focusing system with automatic focusing region selection and point spread function estimation. IEEE Trans. Consum. Electron. **56**(3), 1204–1210 (2010)
12. Sharrab, Y.O., Sarhan, N.J.: Accuracy and power consumption tradeoffs in video rate adaptation for computer vision applications. In: Proceedings of Multimedia and Expo (ICME), pp. 410–415
13. Murino, V., Foresti, G., Regazzoni, C.: Adaptive camera regulation for investigation of real scenes. IEEE Trans. Industr. Electron. **43**(5), 588–600 (1996)
14. Parkany, E., Xie, C.: A complete review of incident detection algorithms and their deployment: what works and what doesn't. IEEE Trans. Industr. Electron. **43**(5), 588–600 (2005)
15. Wan, Y., Huang, Y., Buckles, B.: Camera calibration and vehicle tracking: highway traffic video analytics. Transp. Res. Part C Emerg. Technol. **44**, 202–213 (2014)
16. Grant, C., Gillis, B., Guensler, R.: Collection of vehicle activity data by video detection for use in transportation planning. J. Intell. Transp. Syst. **5**(4), 343–361 (2000)
17. Joshi, A., Atev, S., Fehr, D., Drenner, A., Bodor, R., Masoud, O., Papanikolopoulos, N.: Freeway network traffic detection and monitoring incidents. Cell **136**(2), 215–233 (2007)
18. Preisen, L., Deeter, D.: Next generation traffic data and incident detection from video. Cell **136**(2), 215–233 (2014)
19. Martin, P.T.: Evaluation of UDOT'S video detection systems: system's performance in various test conditions. Cell **136**(2), 215–233 (2004)
20. Shen, M., Kuo, C.: Review of postprocessing techniques for compression artifact removal. J. Vis. Commun. Image Represent. **9**(1), 2–14 (1998)

Index

© Springer International Publishing AG 2017
C. Liu (ed.), *Recent Advances in Intelligent Image Search and Video Retrieval*,
Intelligent Systems Reference Library 121, DOI 10.1007/978-3-319-52081-0

Printed in the United States
By Bookmasters